SEA URCHIN - FROM ENVIRONMENT TO AQUACULTURE AND BIOMEDICINE

Edited by **Maria Agnello**

Sea Urchin - From Environment to Aquaculture and Biomedicine
http://dx.doi.org/10.5772/65503
Edited by Maria Agnello

Contributors

Morgan Ziegenhorn, Davide Sartori, Silvia Giuliani, Andrea Gaion, Simona Macchia, Samantha Lera, David Pellegrini, Lorenzo Morroni, Marina Vaschenko, Peter Zhadan, Tatiana Almyashova, Marta Di Carlo, Maria Di Bernardo, Luciana Dini, Bernardetta Anna Tenuzzo, Elisabetta Carata, Stefania Mariano, Maria Agnello

Published by InTech
Janeza Trdine 9, 51000 Rijeka, Croatia

Publishing Process Manager Martina Usljebrka
Technical Editor SPi Global
Cover InTech Design team

First published October, 2017
Printed in Croatia
Legal deposit, Croatia: National and University Library in Zagreb

Additional hard copies can be obtained from orders@intechopen.com

Sea Urchin - From Environment to Aquaculture and Biomedicine, Edited by Maria Agnello
p. cm.
Print ISBN 978-953-51-3525-8
Online ISBN 978-953-51-3526-5

For EU product safety concerns:
IN TECH d.o.o., Prolaz Marije Krucifikse Kozulić 3, 51000 Rijeka, Croatia,
info@intechopen.com or visit our website at intechopen.com.

Meet the editor

She graduated in Biological Sciences that enabled to the profession of biologist in 2001; in 2005, she achieved the qualification of laboratory technologist and in 2006 the PhD degree in Cellular and Developmental Biology. From 2006 to 2010, she was a research fellow at the current Department STEBICEF of Palermo University, and from 2010 to 2017, she was an adjunct professor of Developmental Biology for the degree course of Biological Sciences. She was an associated research fellow for the SZN of Naples since 2016; in the current year, she achieved the enablement to associate professor for the "Comparative Anatomy and Cytology" sector. Her research activity, focused on stress response, apoptosis, autophagy and mitochondria in embryos and oocytes of sea urchin, has led to several publications on international scientific journals.

Contents

Preface

This book is addressed not only to the readers operating in the sea urchin field of research but also to the lovers of this fascinating organism and refers to interesting investigation on environmental, aquacultural, developmental and biomedical issues. In recent years, the research on sea urchins has been stimulated by the ecological and economic importance of this organism. Whilst several scientists around the world are actively investigating the potential of sea urchin for aquaculture and fisheries, their value as model systems for developmental biology continues to increase. This organism has been elected as model system for the easiness of manipulation, the high amount of available gametes starting from few individuals, the external fertilization and a synchronous embryo development. Since 1880, gametes of sea urchins have been representing a favourite experimental material for studies on fertilization, subsequently for studies on gene activation and regulatory pathways driving embryogenesis. Member of the Echinodermata *phylum*, the sea urchin is among the most known marine invertebrates belonging to deuterostomes. In 2006, the genome of sea urchin *S. purpuratus* has been entirely sequenced, highlighting that this organism is more closely related to humans than other invertebrates, including *D. melanogaster* and *C. elegans*, commonly employed as developmental models. More recent evidences show the validity of sea urchin for studying stress response, apoptosis and autophagy. The sources of environmental stress for a marine organism can be natural or anthropogenic and can alter embryo development, affecting the phenotype. In some case, a hierarchical choice, going from repair to definitive elimination of un-repairable molecules and cells, is activated to face stress. Furthermore, emerging evidences show the suitability of sea urchin embryo for studies of immunology, pathology, toxicology and microbiology.

A collection of selected and relevant researches in the fields of marine, developmental and molecular biology, has contributed to this volume that clearly shows the reason of the high interest manifested by a huge number of scientists around the world for this organism over time. The described topics highlight the validity and versatility of this organism for different kinds of investigations. Each contribution is a separate and comprehensive chapter, but within the book's aim. The book is organized into four sections. The first section contains Introductory chapter: Sea Urchin - Knowledge and Perspectives. The second section about "Environmental Conditions and Sea Urchin Behaviour", contains two interesting chapters. The third section, titled "Aquaculture as a Tool against the Depletion of Sea Urchin Stocks", deepens, through two stimulating contributions, specific aspects of sea urchin aquaculture. As a conclusion of the book, the fourth section is on "Sea Urchin Suitability for Biomedical Studies" and shows new perspectives offered by this model system. In particular, while the first chapter is an Introduction, the second chapter is a comparative review conducted for fifteen urchin genera that use the covering behaviour and suggests a phylogenetic relationship as a functional predictive tool for determining the purpose of covering. The third chapter reports the effects of anthropogenic pressure on the reproduction of *S. intermedius*, showing a shift in timing of spawning closely related to the chronic eutrophication. The fourth chapter concerns the morphological and biochemical profiles of *P. lividus* gonads in natural and breeding conditions. As the environmental conditions have a marked influence on the size, freshness, colour and texture of the gonads, the gonads of wild and breeding male and female were analysed measuring the content of lipids, proteins and carbohydrates for four different types of diet. The fifth chapter reports studies focused on the assessment of a nutritional strategy aimed to ensure, by aquaculture, a gonadal growth of healthy gametes to employ in ecotoxicology. The last chapter of the book highlights the validity of sea urchin embryo as a model for studying the molecular mechanisms involved in aging and age-associated neurodegenerative diseases and as a tool to comprehend the responses to environmental contaminants and chemical compounds.

The validity of sea urchin for molecular, cellular, developmental and eco-toxicological studies is demonstrated by the several published papers. One of the reasons of the ecological importance of this organism for the marine communities is its crucial role in coral reef ecosystems, where it consumes and controls the populations of algae that threaten corals. The economic significance of sea urchin, for fishery and aquaculture, is due to the fact that the edible species, appreciated for the taste and delicacy of the gonads, are part of the diet of humans in many countries. Currently, sea urchin aquaculture is mainly considered to support the human demand, as well as to dispose of a precious remedy against the reduction of biodiversity caused by anthropogenic pollution. Despite the capability of response and adaptation of sea urchin to adverse condition, a strong depletion of natural resources is noticeable; thus, the attention of the scientific community is aimed at the discovery of new strategies, in order to avoid or limit the loss of this important and valid biological resource.

I would like to thank all authors of the accepted chapters for their contribution, the Publishing Process Manager, Ms. Martina Usljebrka,for the precious support in the realization of this book and the InTech Publisher for this stimulating opportunity.

Dr. Maria Agnello, Ph.D.
Università degli Studi di Palermo,
Palermo, Italy

Section 1

Introduction

Chapter 1

Introductory Chapter: Sea Urchin - Knowledge and Perspectives

Maria Agnello

Additional information is available at the end of the chapter

http://dx.doi.org/10.5772/intechopen.70415

1. Introduction

This book is a collection of chapters aimed to provide an overview of some interesting fields of the current research on sea urchin. This magnificent marine world-wide diffused organism offers a myriad of opportunities for investigations on the most various topics, from ecology and toxicology, to aquaculture, as well as to marine, developmental, molecular and cellular biology. To this list of subjects, they must add new emerging biomedical and pharmaceutical applications that, on one side, make this organism even more interesting for those who do not know it from a scientific point of view, and on the other side, rejoins the trust of those who well know its suitability for the basic research, offering further opportunity, e.g. for applied biological investigations.

The sea urchin is an ancient seafloor-dwelling invertebrate, belonging to the phylum of Echinoderms (**Figure 1**), which appeared 520 million years ago, before the Cambrian explosion [1].

Historically, sea urchin has been a key model system in elucidating a variety of classic developmental problems, including fertilization, cleavage, gastrulation and regulation of differentiation in the early embryo [2]. Nowadays, the availability of modern methods of molecular biology allows to deepen specific aspects of the developmental mechanisms, leading to further significant results.

But what are the reasons for the enormous application of this organism for developmental, cellular and molecular studies? These are the most important ones: the gametes can be obtained easily, sterility is not required and the eggs and early embryos of many commonly used species are beautifully transparent. In addition, the early embryo development is highly synchronous. For all these reasons, biochemical and molecular studies result simplified and more feasible.

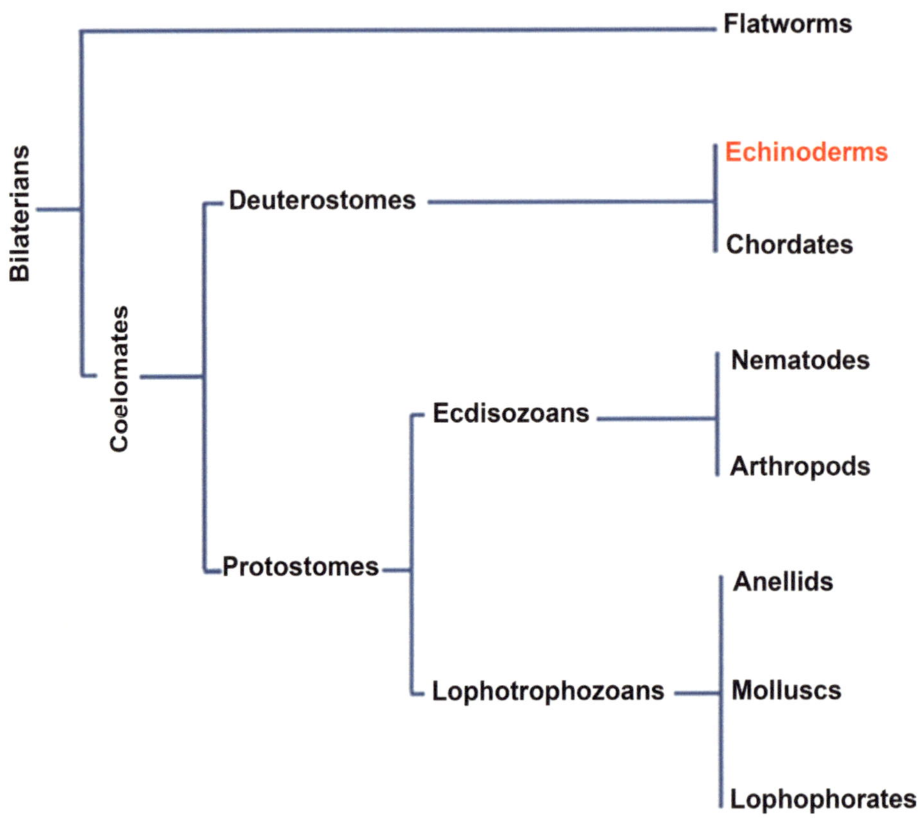

Figure 1. Phylogeny of Echinoderms.

This introductory chapter has been divided in sections, aimed to provide a panorama of the current knowledge and future perspectives offered by using this organism.

2. *Phylum* and ecology

Sea urchin belongs to the class of echinoid, organism of the Echinoderms *phylum*, divided into five main classes: crinoids, asteroids, ophiuroids, echinoids and holothurians (**Figure 2**). Recent molecular studies support the existence of a further class for this *phylum*, that is Asterozoa, an ophiuroid/asteroid clade [3].

Crinoids are considered the most primitive class, whereas echinoids and holothurians are the most advanced. Despite the various shapes, Echinoderms possess common characteristics: benthic lifestyle, adult radial symmetry, water vascular system and calcite endoskeleton with a three-dimensional structure. Their success, testified by the worldwide distribution in several oceanic environments at various depths, latitudes and temperatures, is related to the ability to survive in adverse conditions, by means of strategies such as a spiny defence structure, an immune defence system, a toxin-producing apparatus and an incredible regeneration

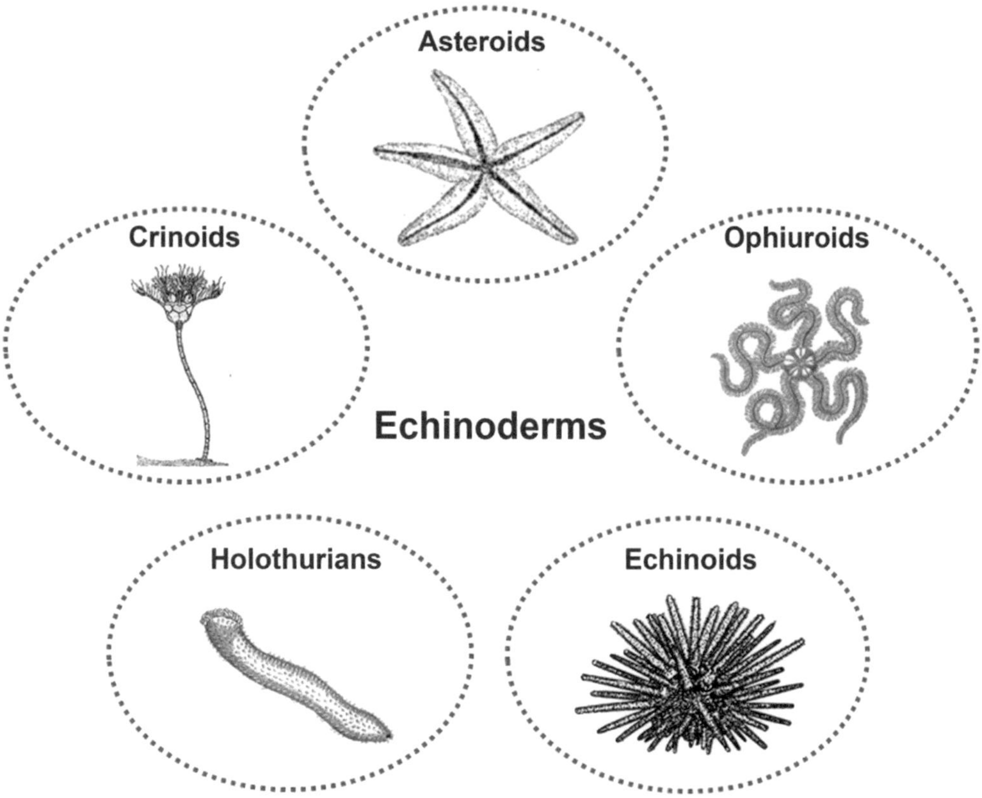

Figure 2. Main classes of the *Phylum* of Echinoderms.

capability [1, 4]. To these strategies, the skill of Echinoderms to face the constant exposure to environmental factors, such as predation, changes in temperature and pH, hypoxia, pathogens, metals, toxicants and pollutants, must be added [5].

Among the five classes of Echinoderms, echinoids are the most various and well represented, from shallow waters to abyssal depths. They are radially symmetrical in adults, ranging from a few millimetres to more than a meter in diameter, and live with their mouth, downward-facing [6]. The group of echinoids includes the sea urchins, heart urchins and sand dollars. Echinoids are usually globular, discoidal or heart-shaped and have a skeleton made of many calcareous plates. Regular echinoids, such as sea urchin, can be distinguished easily from irregular ones by their spherical body, a pentaradial symmetry, and the central location of the anus, above the mouth. The spines are generally long and a jaw apparatus is present in all taxa. Irregular echinoids are elongate in the adult stage. This difference of shape and the posterior position of the anus, instead of dorsally like the regular echinoids, are the two most tell-tale differences setting the two types apart [7, 8].

The spines are used for locomotion, as a passive defence against predators and for trapping floating food particles. Tube feet are located among the spines and are used to move, capture food and attach to substrates. A stinging jaw (or small pincers), called the pedicellariae, is also used for protection and to clutch food items. The mouth of the sea urchin is located on the underside of the organism and consists of a five-pointed jaw called Aristotle's lantern [8].

For their evolutionary position, Echinoderms represent a link between invertebrates and vertebrates, thus occupying a strategic phylogenetic position. This evidence has strongly been sustained by fully sequencing the genome of *Strongylocentrotus purpuratus*: sea urchin is more closely related to humans than other invertebrate models [9, 10].

The echinoderm larval skeleton has been recently proposed as a possible model system for experimental evolutionary biology. It is well known that the evolution of various body plans results from the acquisition of novel structures and/or the loss of existing structures, by multiple and intermediate steps, adaptive or neutral. As the acquisition of a larval skeleton requires multiple steps, this mechanism provides a good model for reproducing intermediate evolutionary stages [11].

Different species of sea urchins live in a variety of marine environments of the world. Some usual places where they live are rock pools and mud, wave-exposed rocks, coral reefs in kelp forests and in sea grass beds, in order to protect themselves from large waves or currents. Sea urchins live in areas where they can find algae, sea grass, seaweed and other foods to consume. One other very important characteristic of these organisms is that they are nocturnal because of their light sensitivity. For this reason, they move spines in reaction to shadows. In order to protect themselves from the powerful ocean currents and waves, sea urchins lodge themselves into holes or crevasses. A common place where it is possible to find a sea urchin is in coral reefs in which it plays a key role in the maintenance of feeding integrity of corals [12]. But sea urchins have also been reported to cause erosion of reefs, so although they are important to the survival of an ecosystem, they can also become dangerous in great numbers [13–15]. In the first chapter of this book, by a comparative analysis among different species of sea urchin, several purposes for the cryptic behaviour are considered: source of food, mechanical defence, protection from sunlight and predators. The results suggest that a phylogenetic relationship provides a functional predictive tool for determining the purpose of the covering.

3. Fishery and aquaculture

Cultivation of sea urchin increasingly becomes a necessity, both for stock enhancement programs and as a means to meet market demand. Sea urchin culture has been practised on a large scale for many decades in Japan, where effective methods have been established. In the rest of the world, sea urchin cultivation is a more recent practice.

It must be consider that when wild stocks decline, the demand created in the market place raises the price of the product and, consequently, culturing is more likely to become economically profitable. Hence the future of the aquaculture industry is closely linked to that of the fisheries, whose fate will ultimately determine the market forces that will shape this growing industry [16]. But it would be desirable that the aquaculture systems would diffuse without waiting for depletion of wild stocks.

There are several examples of how a sea urchin fishery can be managed poorly. The most sobering example was the serial depletion of the Chilean fishery, the main producer and exporter country in the world before 2002. Smaller fisheries exist also in Europe, mainly supplying domestic markets. This fishery has undergone a major transformation to implement management strategies and to avoid a fishery collapse. There is a number of well managed and sustainable sea urchin fisheries around the world that tend to rely on a good overview of the biology of species present in the area as well as sound knowledge of the dynamics of the sea urchin populations. Comprehensive stock assessment and mapping also appears to be an integral part of successful fisheries management [17].

The commercial fisheries and destructive harvesting methods employed to meet market demand, caused, in the last decades, a dramatic depletion of *Paracentrotus lividus*, especially in the Mediterranean, leading to a complete disappearance of urchins from areas of former abundance. Strong aquaculture research programs currently exist in many countries. Several approaches have been tested to respond to the depletion of wild stocks [18–20], but the most challenging strategy could be the setting up of closed aquaculture systems. Several studies have been carried investigating different light/dark regimes, temperature and supply of artificial diets to ensure a rapid gonadal growth and to promote an effective maturation of gametes [21, 22].

In the chapter of Sartori and colleagues the usefulness of formulated diets rich in carotenoids [23–26] was analysed, considering different biological parameters. The authors report the most effective breeding conditions, in Recirculating Aquaculture System, to promote gonadal growth and sexual maturation. In this way, gaps in reproductive events of natural populations can be overcome, always having good quality gametes available for ecotoxicological applications.

Another interesting and cost-effective option is to obtain an uniform quality of gonads, in combination with the systems that use the co-culture with other organisms [27]. As the diet quality can significantly influence the somatic and gonadal growth of sea urchins [28–32], in this book, Tenuzzo and coworkers analyse the morphological and biochemical profiles of the gonadal cycle of *P. lividus,* comparing wild-type organisms with bred ones. These data allow to define the best diet for sea urchins in view of the breeding set-up conditions, representing an important starting point for a scale production of *P. lividus*.

4. Developmental and molecular biology

Sea urchin typically has separate sexes and fertilization takes place externally. During the gametogenic cycle of sea urchin, gonads of both sexes undergo a series of structural changes in charge of the two major cellular populations of the germinal epithelium: (i) germinal cells, i.e. oogonia or spermatogonia and (ii) somatic cells, called nutritive phagocytes, present in both sexes, in which nutrient storage occurs. Nutritive phagocytes are somatic cells that provide a structural and nutritional microenvironment for germinal cells during gametogenesis [33]. The oocytes of sea urchin, differently from other Echinoderms and other animals, complete meiotic maturation before fertilization. During oogenesis, several processes occur to ensure the correct formation of the mature eggs. Recent data have shown that the autophagic process is required for sea urchin oogenesis undoubtedly as a catabolic basal mechanism. The major concentration of vesicular acidic organelles is near the germinal vesicle. Interestingly, the mitochondria are more active in the same regions [34, 35]. The manifestation in the same areas of two catabolic mechanisms, oxidative phosphorylation and autophagy, is interesting as their coexistence could occur to ensure the final steps of the meiotic process.

Spawning of gametes occurs in the spring and early summer, when water and food conditions are favourable. Females will release several million eggs into the water which will unite with sperms released from the males. Embryo development, till pluteus stage, can be followed continuously in real time by microscopy, because the embryos are optically transparent and no histological preparation is required.

After hatching and gastrulation, the free swimming larvae, with a planktonic existence which lasts several months, undergo metamorphosis, before they settle to the bottom. Sexual maturity is reached at 4–10 years, with a diameter of 1.8–2.5 cm. As sea urchin embryo develops rapidly (15- to 60-min cell cycles depending on the species), synchronously and predictably, large cultures can be cultivated, treated and dissociated, while individual embryos can be injected or manipulated to test important features of their mechanisms of development. The micromeres obtained after the fourth cell division are the major signalling centre of the embryo [2, 36]. The mechanism behind this induction is poorly understood, but the micromeres do acquire nuclear β-catenin shortly after their formation, and evidence suggests Notch-Delta and Wnt8 signalling as candidates for this inductive process [37]. Perturbation of the transcription factors and signals provided the means for assembling models of the gene regulatory networks used for specification and the control of the subsequent morphogenetic events. By this tool, it was learned that ectoderm provides a series of patterning signals to the skeletogenic cells and as a consequence the skeletogenic cells secrete a highly patterned skeleton based on their ability to genotypically and locally decode several signals [38].

The ability to trace the entire trajectory of specification of a cell type until it terminally differentiates is now a realistic goal. In its current form, the sea urchin genes regulatory network includes more than 100 transcription factors and a number of signalling pathways, and in most cases, multiple laboratories have validated each connection in different sea urchin species, such as *Lytechinus* and *Paracentrotus*, starting from *S. purpuratus*. Using a similar tool, it is now possible to analyse and understand how a cell type arises and works in detail. This will be valuable in understanding the entire history of neurogenesis, and in determining the mechanisms by which neurons diversify toward different neurotransmitter cell types [39].

5. Defence mechanisms: a success guarantee!

Sea urchin is considered a good bioindicator of marine health, as well as a good tool for testing the toxicity of environmental and chemical molecules. Currently, it is one of the organism considered by the European Agency for Alternative models and in full respect of the 3Rs objectives (reduction, refinement and replacement of animal experiments); thus, it could be a suitable organism for toxicological studies as alternative to vertebrates. Sea urchin is able to detect both biotic and abiotic stress and to recognize, transform and eliminate many potentially harmful materials to protect itself.

The immune defence of sea urchin, appeared early in the evolution, is mediated by a vast repertoire of molecules [5]. Human require both adaptive and innate immune responses, whereas sea urchin require only innate immune functions, as revealed by the several innate immunity gene families discovered [40]. The immune cells from the adult *P. lividus*, which have been introduced in the past 15 years as sentinels of environmental stress and for toxicological testing, are valid tools to discover basic molecular and regulatory mechanisms of immune responses [5]. Echinoderm immune cells, also known as coelomocytes, are a heterogeneous population of freely moving cells found in all coelomic spaces, including the perivisceral cavities and the water vascular system. They are also present sparsely in the connective tissue and among tissues of various organs [41–44]. A rapid increase of the number of the red amoebocytes has been shown in samples taken from polluted sea water, whereas petaloid cells are actively involved in the phagocytosis and phylopodial cells trigger the clot formation [5].

A recent and promising search path concerns the antimicrobial peptides (AMPs), important effector molecules in innate immunity. Echinoderms live in a microbe-rich marine environment and are known to express a wide range of AMPs. Recently, two novel AMP families from coelomocytes of sea urchins have been found: strongylocins and centrocins [45].

Another defence strategy employed by some species of sea urchin is represented by toxins. The spines of the species *Tripneustes gratilla* are considered toxin-producing organs. Upon being stung by this sea urchin, humans and animals could be poisoned, with difficulty breathing, muscle paralysis, convulsion and other symptoms [46].

Several toxicological studies analyse the possibility that a number of environmental molecules and xenobiotics interfere with embryogenesis and cause developmental anomalies [47–49]. The effects of environmental anthropogenic factors on reproduction of the sea urchin *Staphylococcus intermedius* are considered in the second chapter of this book. The results of long-term studies in wild populations, located in the north-western Sea of Japan and with different levels of anthropogenic pressure, indicate that the principal external factor determining a shift in timing of spawning is the phytoplankton concentration, closely related to the chronic eutrophication.

Environmental stress usually causes severe effects on embryogenesis, but sea urchin is able to detect stressful insults and to activate a response. It is well known that embryos of different species, exposed to different kind of stress, temporarily suspend their development and activate several protective strategies, including programmed cell death. Sea urchin activates several cellular strategies of defence, such as synthesis of heat shock proteins, apoptosis and autophagy, to face different stress and survive in adverse conditions. The protective action of heat shock proteins is not always sufficient to block the toxic action of a stressor. When the cell damage is too great, mechanisms of programmed cell death, such as apoptosis and autophagy, may occur and contribute to the elimination of the irreversibly damaged cells in order to maintain the integrity of the embryos (**Figure 3**).

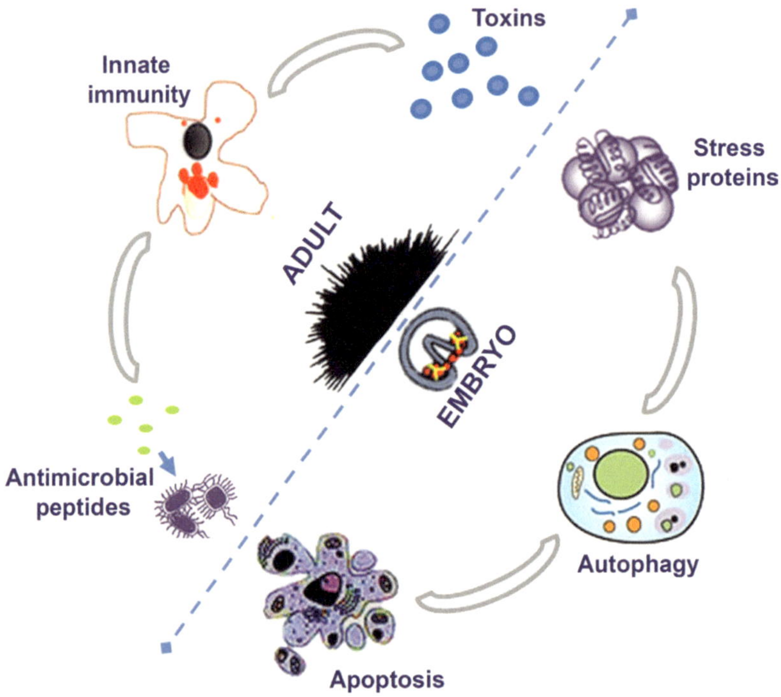

Figure 3. Main defence strategies of sea urchin adults and embryos.

A xenobiotic able to cause a strong activation of all embryonic defence mechanisms reported in (**Figure 3**) is cadmium, an heavy metal [47, 50–53]. It must be underlined that such defence mechanisms are basally activated in physiological conditions and that their extent highly increases in stress conditions.

In the last chapter of this book, the authors draw attention on the validity of sea urchin embryo as a model to test the developmental effects and toxicity responses against various environmental contaminants and chemical compounds.

The evaluation of the defensive responses of sea urchin is of high interest to researchers and marine organizations for the management of biological resources and ecosystems [54].

6. New perspectives

Like many other marine organisms, Echinoderms have been, and continue to be, examined as a source of biologically active compounds with biomedical applications [16]. Multiple species of sea urchin were used anciently in the traditional Chinese medicine for treating several diseases. The shells of yellow sea urchin, everyday consumed, are discarded, although they could be employed by pharmacologists for discovering the medicinal components. Several examples of Echinoderms derived substances that may have biotechnological value and therapeutic application have been reported, e.g. antibiotic, antifouling and anticancer molecules [55]. The major chemical and bioactive constituents of sea urchin have recently been summarized: pigments, proteins, toxins, polysaccharide, minerals, sterols, vitamins, fatty acids and amino acids. Modern medical studies have proved that shells, spines, gonads and other parts of sea urchins have a good medicinal value. Ulcer disease, otitis media, cervical lymph node tuberculosis and pain are some examples of pharmaceutical applications. The toxic, haemolytic and hypotensive functions of white sea urchin spines have been used to develop neuromuscular blocking drugs [56]. For all these reasons, sea urchin clearly represents a fairly unexploited source in the identification of new and useful drugs. These bioactive molecules were produced in the course of millions of years of evolution in order to protect the individual, the cells or the entire organism from possible competitors, are continuously evolving molecules that interfere with signalling or are simply toxic for other cells [57]. A novel and promising search path concerns antimicrobial peptides (AMPs). The echinodermal AMPs are crucial molecules for the understanding of immunity, and their potent antimicrobial activity makes them potential precursors of novel drugs [45].

In 2006, it has been shown that 7077 sea urchin genes are conserved in humans, and many of these may correlate with cancer, Huntington's disease, muscular dystrophy, atherosclerosis and other illnesses [9]. The last chapter of this book offers an interesting overview of the contribution of sea urchin embryo as a simple model to study ageing and age-associated neurodegenerative diseases, as well as the pathways involved in cell survival and death.

7. Conclusions

Since the beginning of cell biology, the Echinoderms have been employed as embryological model systems to study basic phenomena, such as mitosis, cell division, differentiation and organ formation, by the famous cell biologists in the nineteenth and twentieth centuries, such as Boveri, Heilbrunn, Mazia, Monroy, Wilson, Hertwig and Brachet. Today, sea urchin continues to be the model of choice for several cellular and molecular biologists, contributing to significant advance in the fields of evolutionary and developmental biology and leading to the discovery of complex gene regulatory networks. In the same time, its usefulness as a sensitive system for toxicological test and environmental studies continues to increase.

Although at present, sea urchins seem to be in no immediate danger of disappearing or becoming endangered; in general, they have shown mass mortality caused by an increased pollution in the marine environment, an increased amount of fishing by humans, as well as a rise in the temperature of water due to global warming. As sea urchins are very susceptible to changes, they may become endangered in the future; thus, the use of aquaculture systems is desirable.

Furthermore considering the current interest for novel drugs discovery, the urgent need for novel antibiotics against the bacterial resistance and the need of new strategies in cancer research, sea urchin represents an important resource to exploit in the biotechnology and medicine field.

Acknowledgements

A special thanks to Prof. Maria Carmela Roccheri and Anna Maria Rinaldi, the main guides in my research path on sea urchin.

Author details

Maria Agnello[1,2]

Address all correspondence to: maria.agnello@unipa.it

1 Dipartimento di Scienze e Tecnologie Biologiche, Chimiche e Farmaceutiche, Università degli Studi di Palermo, Palermo, Italy

2 Dipartimento di Biologia ed Evoluzione Organismi Marini, Stazione Zoologica "Anton Dohrn," Napoli, Italy

References

[1] Bottjer DJ, Davidson EH, Peterson KJ, Cameron RA. Paleogenomics of Echinoderms. Science 2006;314:956-960

[2] Hörstradius S. The mechanics of sea urchin development, studied by operative methods. Biological Reviews of the Cambridge Philosophical Society 1939;14:132-179

[3] Telford MJ, Lowe CJ, Cameron CB, Ortega-Martinez O, Aronowicz J, Oliveri P, et al. Phylogenomic analysis of echinoderm class relationships supports Asterozoa. Proceedings of the Biological Sciences 2014;281:1786

[4] Iken K, Konar B, Benedetti-Cecchi L, Cruz-Motta JJ, Knowlton A, Pohle G, et al. Large-scale spatial distribution patterns of Echinoderms in nearshore rocky habitats. PLoS One 2010;5:e13845

[5] Pinsino A, Matranga V. Sea urchin immune cells as sentinels of environmental stress. Developmental and Comparative Immunology. 2015;49(1):198-205

[6] Sreepat J. Fundamentals of invertebrate palaeontology: Macrofossils. Chapter 6. Springer Geology; India. 2017. pp. 175-209. ISBN: 9788132236580

[7] Smith AB and Kroh A. Phylogeny of sea urchin. Chapter 1. In: Lawrence JM editor. Sea Urchins: Biology and Ecology. 3rd ed. Elsevier Academic Press; United Kingdom. 2013. pp. 1-12. ISBN: 9780123964915

[8] Ruppert EE, Fox R, Barnes RD. Invertebrate Zoology: A Functional Evolutionary Approach. Thomson-Brooks/Cole, United States. 2004. pp. 905. ISBN: 9780030259821

[9] Sodergren E, Weinstock GM, Davidson EH et al. The genome of the sea urchin *Strongylocentrotus purpuratus*. Science 2006;314:941-952

[10] Davidson EH. The sea urchin genome, where would it lead us? Science 2006;314:939-940

[11] Koga H, Morino Y, Wada H. The echinoderm larval skeleton as a possible model system for experimental evolutionary biology. Genesis 2014;52:186-192

[12] Hughes TP, Reed DC, Boyle MJ. Herbivory on coral reefs: Community structure following mass mortalities of sea urchins. Journal of Experimental Marine Biology and Ecology. 1987;113(1):39-59

[13] Ralph B, Buchsbaum M, Pearse J, Pearse V. Animals without Backbones. 3rd ed. London and Chicago: The University of Chicago; 1987. pp. 478-482. ISBN: 9780226078748

[14] David G, and George J. Marine Life: An Illustrated Encyclopaedia of Invertebrates in the Sea. New York: A Wiley-Interscience Publication; 1979. pp. 243

[15] Sue W, Hanna N. The Greenpeace Book of Coral Reefs. New York: Sterling Co., Inc; 1992. pp. 64-65

[16] Kelly MS. Echinoderms: their culture and bioactive compounds. Progress in Molecular and Subcellular Biology 2005;39:139-165

[17] James P, Noble C, Hannon C, Stefánsson G, Þórarinsdóttir G, Sloane R, Ziemer N, Lochead J. Sea urchin fisheries, management and policy review (Activity A4.2.1 of the URCHIN project); 2016. Nofima Report 18/2016

[18] Yokota Y. Fishery and consumption of the sea urchin in Japan. In: Yokota Y, Matranga V, Smolenicka Z, editors. The Sea Urchin: From Basic Biology to Aquaculture. Lisse, The Netherlands: Swets & Zeitlinger BV; 2002. pp. 129-138

[19] Cook EJ, Kelly MS. Enhanced production of the sea urchin *Paracentrotus lividus* in integrated open-water cultivation with Atlantic salmon *Salmo salar*. Aquaculture 2007; 273:573-585

[20] Pantazis PA. The culture potential of *Paracentrotus lividus* (Lamarck 1816) in Greece: a preliminary report. Aquaculture International 2009;17:545-552

[21] Grosjean P, Spirlet C, Gosselin P, Vaitilingon D, Jangoux M. Land-based, closed-cycle echiniculture of *Paracentrotus lividus* (Lamarck) (Echinodermata: Echinoidea): A long-term experiment at a pilot scale. Journal of Shellfish Research 1998;17:1523-1531

[22] Devin MG. Land-based echinoculture: A novel system to culture adult sea urchins. In: Yokota Y, Matranga V, Smolenicka Z, editors. The Sea Urchin: From Basic Biology to Aquaculture. Lisse, The Netherlands: Swets & Zeitlinger BV; 2002. pp. 145-159

[23] Vadas RL, Beal B, Dowling T, Fegley JC. Experimental field tests of natural algal diets on gonad index and quality in the green sea urchin, *Strongylocentrotus droebachiensis*: A case for rapid summer production in post-spawned animals. Aquaculture 2000;182:115-135

[24] Bendich A. Recent advances in clinical research involving carotenoids. Pure and Applied Chemistry 1994;66:1017-1024

[25] Tsushima M, Kawakami T, Mine M, Matsuno T. The role of carotenoids in development in sea urchin *Pseudocentrotus depressus*. Invertebrate Reproduction & Development 1997;32(2):149-153

[26] Kawakami T, Tsushima M, Katabami Y, Mine M, Ishida A, Matsuno T. Effect of β,β-carotene, β-echinenone, astaxanthin, fucoxanthin, vitamin A and vitamin E on the biological defense of the sea urchin *Pseudocentrotus depressus*. Journal of Experimental Marine Biology and Ecology 1998;226:165-174

[27] Cook EJ, Kelly MS. Co-culture of the sea urchin *Paracentrotus lividus* and the edible mussel *Mytilus edulis* L. on the west coast of Scotland, United Kingdom. Journal of Shellfish Research. 2009;28:553-559

[28] Schlosser SC, Lupatsch I, Lawrence JM, Lawrence AL, Shpigel M. Protein and energy digestibility and gonad development of the European sea urchin (Lamarck) fed algal and prepared diets during spring and fall. Aquaculture Research 2005;36:972-982

[29] Meidel SK, Scheibling RE. Effects of food type and ration on reproductive maturation and growth of the sea urchin *Strongylocentrotus droebachiensis*. Marine Biology1999;134:155-166

[30] Otero-Villanueva MDM, Kelly MS, Burnell G. How prey size, type and abundance affects foraging decisions by the regular echinoid *Psammechinus miliaris*. Journal of the Marine Biological Association of the United Kingdom 2006;86:773-781

[31] Haya D, Régis MB. Comportement trophiqué de *Paracentrotus lividus* (Lam.) (Echinodermata: Echinoidea) soumis à six regimés alimentaires dans des conditions éxperimentales. Mesogée.1995;54:35-42

[32] Chopin T, Bushmann AH, Halling C, Troell M, Kautsky N, Neori A. Integrating seaweeds into marine aquaculture systems: A key toward sustainability. Journal of Phycology 2001;37:975-986

[33] Walker CW, Lesser MP and Unuma T. Sea urchin gametogenesis—Structural, functional and molecular/genomic biology. Chapter 3. In: Lawrence JM, editor. Sea Urchins: Biology and Ecology; 3rd ed. Elsevier Academic Press; United Kingdom. 2013. pp. 25-43. ISBN: 9780123964915

[34] Agnello M, Chiarelli R, Martino C, Bosco L, Roccheri MC. Autophagy is required for sea urchin oogenesis and early development. Zygote. 2016;24(6):918-926

[35] Agnello M, Roccheri MC, Morici G, Rinaldi AM. Mitochondria during sea urchin oogenesis. Zygote. 2017;25(2):205-214

[36] Ransick A, Davidson EH. A complete second gut induced by transplanted micromeres in the sea urchin embryo. Science. 1993;259(5098):1134-1138

[37] McClay DR. Evolutionary crossroads in developmental biology: Sea urchins. Development. 2011;138(13):2639-2648

[38] McClay DR. Sea urchin morphogenesis. Current Topics in Developmental Biology 2016;117:15-29.

[39] Martik ML, Lyons DC, McClay DR. Developmental gene regulatory networks in sea urchins and what we can learn from them. pii: F1000 Faculty Rev-203. eCollection 2016. DOI: 10.12688/f1000research.7381.1

[40] Smith LC. Diversification of innate immune genes: lessons from the purple sea urchin. Disease Models & Mechanisms 2010;3(5-6):274-279

[41] Glinski Z. Jarosz J. Immune phenomena in Echinoderms. Archivum Immunologiae et Therapiae Experimentalis 2000;48:189-193

[42] Smith LC, Ghosh J, Buckley KM, Clow LA, Dheilly NM, Haug T, et al. Echinoderm immunity. Advances in Experimental Medicine and Biology 2010;708:260-301

[43] Munõz-Chápuli R., Carmona R, Guadix JA, Macías D, Pérez-Pomares JM. The origin of the endothelial cells: An evo-devo approach for the invertebrate/vertebrate transition of the circulatory system. Evolution and Development 2005;7:351-358

[44] Pinsino A, Thorndyke MC, Matranga V. Coelomocytes and post-traumatic response in the common sea star *Asterias rubens*. Cell Stress & Chaperones 2007;12:331-341

[45] Li C, Blencke HM, Haug T, Stensvåg K. Antimicrobial peptides in echinoderm host defense. Developmental and Comparative Immunology 2015;49(1):190-197

[46] Wang J, Gong XG. Research progress on anti-tumor and immune regulation of polysaccharide. Chinese Journal of Biochemistry and Pharmacology. 2001;22(1):52-54

[47] Roccheri MC, Agnello M, Bonaventura R, Matranga V. Cadmium induces the expression of specific stress proteins in sea urchin embryos. Biochemical and Biophysical Research Communications 2004;321(1):80-87

[48] Pinsino A, Matranga V, Trinchella F, Roccheri MC. Sea urchin embryos as an in vivo model for the assessment of manganese toxicity: Developmental and stress response effects. Ecotoxicology. 2010;19(3):555-562

[49] Falugi C, Lammerding-Koppel M, Aluigi MG. Sea urchin development: An alternative model for mechanistic understanding of neurodevelopment and neurotoxicity. Birth Defects Research. Part C, Embryo Today. 2008;84(3):188-203

[50] Agnello M, Filosto S, Scudiero R, Rinaldi AM, Roccheri MC. Cadmium induces an apoptotic response in sea urchin embryos. Cell Stress & Chaperones. 2007;12(1):44-50

[51] Agnello M, Roccheri MC. Apoptosis: Focus on sea urchin development. Apoptosis. 2010;15(3):322-330

[52] Chiarelli R, Agnello M, Roccheri MC. Sea urchin embryos as a model system for studying autophagy induced by cadmium stress. Autophagy. 2011;7(9):1028-1034

[53] Agnello M, Bosco L, Chiarelli R, Martino C and Roccheri MC. The role of autophagy and apoptosis during embryo development. In: Ntuli TM, editor. Cell Death—Autophagy, Apoptosis and Necrosis. Rijeka, Croatia: InTech; 2015. pp. 84-112. ISBN: 978-953-51-2236-4

[54] Chan F, Boehm AB, Barth JA, et al. The West Coast Ocean Acidification and Hypoxia Science Panel: Major Findings, Recommendations, and Actions. Oakland, California, USA: California Ocean Science Trust; 2016

[55] Petzelt C. Are Echinoderms of interest to biotechnology? Progress in Molecular and Subcellular Biology 2005;39:1-4

[56] Shang XH, Liu XY, Zhang JP, Gao Y, Jiao BH, Zheng H, Lu XL. Traditional Chinese medicine—Sea urchin. Mini-Reviews in Medicinal Chemistry. 2014;14(6):537-542

[57] Haug T, Kjuul AK, Styrvold OB, Sandsdalen E, Olsen ØM, Stensvåg K. Antibacterial activity in *Strongylocentrotus droebachiensis* (Echinoidea), *Cucumaria frondosa* (Holothuroidea), and *Asterias rubens* (Asteroidea). Journal of Invertebrate Pathology. 2002;81(2):94-102

Section 2

Environmental Conditions and Sea Urchin Behaviour

Chapter 2

Sea Urchin Covering Behavior: A Comparative Review

Morgan A. Ziegenhorn

Additional information is available at the end of the chapter

http://dx.doi.org/10.5772/intechopen.68469

Abstract

Covering behavior in sea urchins is an important aspect of many species' ecology and has a variety of perceived benefits including food source, mechanical defense, shielding from sunlight, and predator protection. The goal of this study was to determine whether an urchin genus's main benefit from this form of crypsis is correlated with either phylogenetic relationships or environmental factors (ocean depth and climate). To evaluate this hypothesis, a literature review was conducted on 15 urchin genera that use the covering reaction. The function of this behavior for the aforementioned genera was both mapped onto a phylogeny and evaluated, based on the climate and depth of the genera's habitats to determine whether the patterns exist. The results suggest that phylogenetic relationships provide a more functional predictive tool for determining the purpose of covering in an urchin genus than its environment. This conclusion is useful for understanding the biology of sea urchins as well as how the covering reaction relates to the many other cryptic behaviors used by animal species.

Keywords: crypsis, ecology, protective behavior, phylogenetics

1. Introduction

The crucial roles that sea urchins play in aquaculture and coral reef ecosystems warrant the study of their ecology and behavior [1, 2]. Many species exhibit a special form of crypsis called covering behavior [3–5], also known as "dressing" or "heaping" [6] (**Figure 1**), which involves securing algae, rocks, and other items in their environment to their tests using a combination of spines and tube feet [7]. While the mechanics of this reaction are well understood, the reasons for it remain nebulous.

Two conflicting hypotheses exist with respect to this form of crypsis: the "reflexive" hypothesis, which proposes that the covering reaction is automatic, and the "functional" hypothesis, which suggests that heaping is used for specific purposes. The reflexive hypothesis has been supported

Figure 1. A covered sea urchin. An example of the covering behavior of the sea urchin *Tripneustes gratilla* in Hawaii, USA.

by several studies. One such study examined the covering behavior of *Strongylocentrotus* urchins and concluded that the reaction is innate because it is not affected by inbreeding or urchin age [8], suggesting that covering might not be decision based. It has also been noted that the stimuli often seemed to promote more activity in the animal [9]. Another study argued that the movements involved in moving covering material from one part of the test to another were indistinguishable from the urchin's movement when righting itself if flipped over, which is reflexive [10]. Based on these results, these studies concluded that covering is merely a by-product of additional movement.

In contrast, the functional hypothesis theorizes that covering behavior is a conscious decision by sea urchins. Many species have been shown to choose when to cover and to what extent, based on a variety of environmental stimuli [4, 5, 7, 11–13]. One study noted the differences between both species and individuals in the amount of covering even in similar environmental conditions, which suggest that this crypsis is a decision, as otherwise all urchins would cover themselves to an equal extent [14]. Additional research has found that urchins choose objects of particular sizes with which to cover themselves [15], a result that would not be expected were the action reflexive. While both hypotheses have convincing evidence, the functional hypothesis has been assumed for the current study based on the variety of species-specific reactions to environmental stimuli, in addition to the fact that not all urchin species use heaping behavior.

Covering can provide sea urchins with protection from predators [16], mechanical damage (wave surge or floating debris) [11, 17], overexposure to light [7], or can act as a food source [18]. However, no prior studies have considered a relationship between sea urchin biology and the reaction. Such a relationship would provide both a framework to understand why such a variety of applications for covering have been observed and a tool for predicting what function crypsis serves in an unstudied species of urchin.

http://dx.doi.org/10.5772/intechopen.68469

One source for such a framework might be environmental considerations. The reason for crypsis and the extent to which it is seen differs widely between species [14]. However, the behavior is also affected by size and mobility [19] as well as the energetic cost of covering [20]. It is possible, then, that the differences in the covering reaction between species are due to variations in animal size and mobility as well as energetic costs, rather than genetics. In this case, the reason for covering might be deduced not by a species' evolutionary relationships, but rather by the climate and ocean depth at which it is found.

It has been shown, though, that differences in the heaping reaction persist even for species such as *Strongylocentrotus purpuratus* and *Paracentrotus lividus* that share very similar environments [21]. This points to the possibility that covering, and perhaps, a particular genera's reason for using the behavior, has an evolutionary basis. The covering behavior of some sea urchin genera has been previously mapped onto a phylogeny of sea urchin species which demonstrated that it likely evolved once and was subsequently lost from the genera *Glyptocyphus* and in the *Colobocentrotus-Heliocidaris* clade [5] (**Figure 2**). The finding that species exhibiting heaping are closely related suggests that some genetic basis does exist.

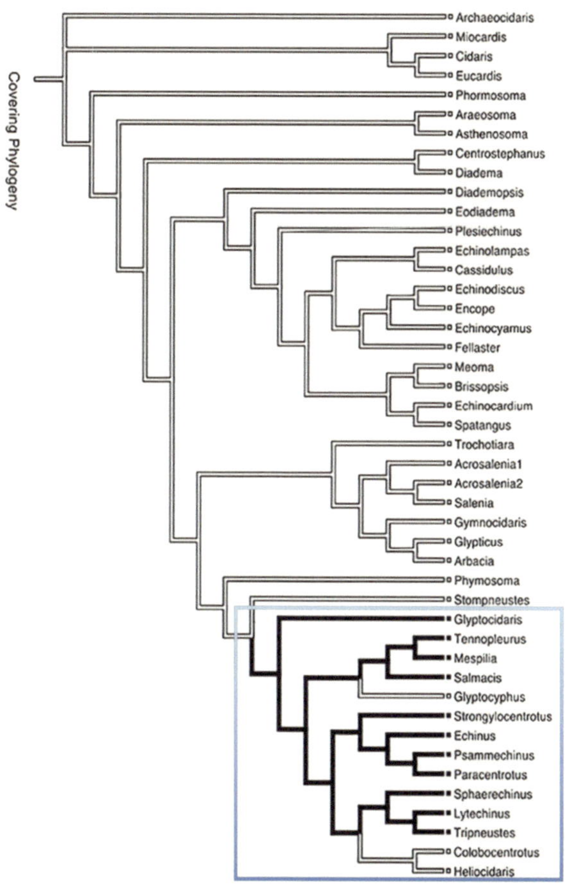

Figure 2. Sea urchin phylogeny. A phylogeny of urchin genera courtesy of Ziegenhorn (2016). The species that cover are noted in dark lines within the box.

The purpose of this study was to review all existing studies of cryptic sea urchins to determine whether covering behavior can be predicted from either (1) the depth and climate within which an urchin resides (the species' environment), or (2) the species' relationship to other cryptic sea urchins.

2. Methods

Studies of 15 urchin genera were researched to determine the reason for their cryptic behavior, chosen based on a previous paper that researched which species of sea urchins use covering [5]. The genera considered were *Glyptocidaris, Temnopleurus, Mespilia, Salmacis, Strongylocentrotus, Echinus, Psammechinus, Paracentrotus, Sphaerechinus, Lytechinus, Tripneustes, Pseudoboletia, Toxopneustes, Pseudechinus,* and *Genocidaris.*

Four main reasons for covering were considered: predator defense, protection from mechanical damage (wave surge/floating debris), use as a food source, and protection from bright light (sunlight/UV). When it was possible, this list was narrowed down to one main use for crypsis, though in some cases, two uses had equal evidence and were both noted as main covering reasons. For some genera (*Pseudoboletia, Genocidaris, Psammechinus, Mespilia,* and *Temnopleurus*), it was impossible to determine the reason for covering behavior. In this case, the explanation was considered unknown.

2.1. Phylogenetic methods

To determine whether evolutionary relationships provide a solid framework for urchin's covering motivation, the use of heaping by eleven of the 15 urchin genera was mapped onto an existing phylogeny [5] (**Figure 2**). The genera included in the phylogeny were *Glyptocidaris, Temnopleurus, Mespilia, Salmacis, Strongylocentrotus, Echinus, Psammechinus, Paracentrotus, Sphaerechinus, Lytechinus,* and *Tripneustes.*

2.2. Environmental condition methods

To consider whether the environmental conditions are the best indicator of covering purposes, an additional four urchin genera for which the reaction has been noted (*Pseudoboletia, Toxopneustes, Pseudechinus,* and *Genocidaris*) were reviewed. A most likely reason for crypsis in these four genera was assigned and information on the depth and area of the world in which all fifteen genera have been found was determined from a previous study [22]. Because of its wide range of climates and ocean depths, the genus *Strongylocentrotus* was divided into three species (*S. purpuratus, Strongylocentrotus intermedius,* and *Strongylocentrotus drobachiensis*) that each had different climate and depth ranges. These data, combined with the most likely explanation for covering behavior, were used to create two tables: one for ocean depth and one for ocean climate. Ocean depth ranges were divided into five categories: 0–50 , 0–100, 0–250, 0–1000, and 0–1200 meters. Ocean climate ranges were divided into four categories: tropical, tropical/temperate, temperate, and temperate/polar. Two additional maps were created to illustrate the range of each species with regard to various sea surface temperatures (SST), and this map was modified to show the different uses of heaping in regards to SST.

3. Results

3.1. Covering as sunlight/UV protection

Protection from sunlight was the most common reason for using the behavior in the urchin genera considered, with six of the total fifteen genera covering primarily for this reason. *Lytechinus* urchins tested in the lab react to UV light and direct sunlight, but discard covering in darkness, exhibiting a diurnal pattern of crypsis [15, 23–25]. Because urchins discard covering in darkness, but tropical urchins are often nocturnal, it was concluded that covering was likely not a source of shelter from predation [25]. This study also found that urchins covered in response to wave surge regardless of the light condition [25]. However, other studies of *Lytechinus* support the hypothesis that their cryptic behavior is primarily light-based as it is affected by light in the absence of wave surge action [15, 25]. In addition to this, one study noted that covering might even be a mechanical disadvantage, as it makes it more difficult for the urchin to right itself, and there is no evidence that heaping provides additional mechanical defense [25].

For *Tripneustes* urchins, it has been found that albino urchins cover more than pigmented specimens and that both albino and pigmented *Tripneustes* urchins heap more in response to sunlight in lab settings [5, 26, 27]. In the field, *Tripneustes* urchins cover more when exposed to light [27] and show no preference for being under rocks that might better protect them from predators and wave surge [5]. Additionally, *Tripneustes* urchins do not prefer either algal or coral materials, which suggest that cover is not an important food source and is not used to weigh urchins down [5].

Several other urchin genera also likely cover to shelter themselves from overexposure to sunlight. *Salmacis* urchins have been found to heap significantly more in response to light, particularly UV-C [12]. Light protection also remains the dominant explanation for the *Paracentrotus* heaping reaction, and lab studies of covering have shown that *Paracentrotus* urchins avoid bright light if possible [4, 28]. It has also been suggested that the behavior has little to do with predators, as *Paracentrotus* urchins have been observed to cover most in the summer when their predators are least active [4, 28]. Some studies have considered floating debris and food source as factors that affect the covering reaction [3, 11], but in both studies, the effect of light was not considered. For *Sphaerechinus* urchins, an ability to cover and a sensitivity to UV light have been noted, which makes it likely that light is the reason for their cryptic behavior [29, 30]. For *Strongylocentrotus* urchins, UV light seems to be one of two main reasons for the reaction along with using covering materials as a food source. Previous studies of this genera have suggested that the urchins in the lab heap more in response to UV-A and UV-B rays [7], and that males in particular show a strong response when exposed to higher intensities of sunlight [31].

3.2. Covering as mechanical protection

Of the urchin genera considered, two were found to cover primarily for mechanical protection (*Glyptocidaris* and *Toxopneustes*). *Glyptocidaris* urchins show increased cryptic behavior in the lab when sand is poured into their environment, which suggests that these urchins cover themselves to shield from floating debris [13]. Another study noted that starved *Glyptocidaris* urchins continue to cover themselves, which suggests that the behavior is important for survival, as it persists even when the energetic cost of maintaining it is fairly high [32]. *Toxopneustes* urchins have been observed in the field covering themselves to differing extents in relation to their body size, as if to achieve a certain heft that keeps them secure on rocks [33]. This same study also noted that urchins in areas of highest current have a higher covering to body weight ratio and concluded from these observations that these urchins use the reaction to defend themselves from wave surge [33].

3.3. Covering as a food source

Two urchin genera, *Echinus* and *Strongylocentrotus,* use covering as a food source. *Echinus* sea urchins have been observed using primarily algae species that are a part of their diet as cryptic material, and from this, it has been concluded that heaping is likely a way of capturing food, as these urchins are deep enough in the ocean (up to 1200 meters) that visible light or hydro-dynamics seem less likely explanations for the reaction [34, 35]. *S. purpuratus* urchins do not show a diurnal pattern of covering, and, though they seem to cover in areas with more wave surge, their behavior is most likely linked to food capture as their hold on the rocks is strong enough to defend them from strong currents [18, 36].

3.4. Covering as predator protection

For *Pseudechinus* urchins, cryptic behavior seems to be an effective method of predator defense in the field [16]. One study of *Strongylocentrotus* urchins found that covering can protect urchins in the lab setting from predation by crabs for short periods of time, up to 2 hours [1]. However, a more recent experimental study of these urchins found no correlation between amount of covering and presence of predators, suggesting that protection of this type is coincidental rather than a functional reason for heaping in this genera [37].

3.5. Covering for unknown reasons

While both *Pseudoboletia* and *Genocidaris* urchins have shown crypsis [38, 39], no information was found regarding the reasons for this behavior in either genera. For *Psammechinus, Mespilia,* and *Temnopleurus* urchins, phototaxis has been noted but the covering reaction has not been experimentally studied [6, 7, 40–42].

3.6. Covering by evolutionary relationship

The various reasons for urchin heaping were mapped on to the existing phylogenetic tree (**Figure 3**). As previously noted, *Lytechinus, Tripneustes,* and *Sphaerechinus* urchins all cover to

http://dx.doi.org/10.5772/intechopen.68469

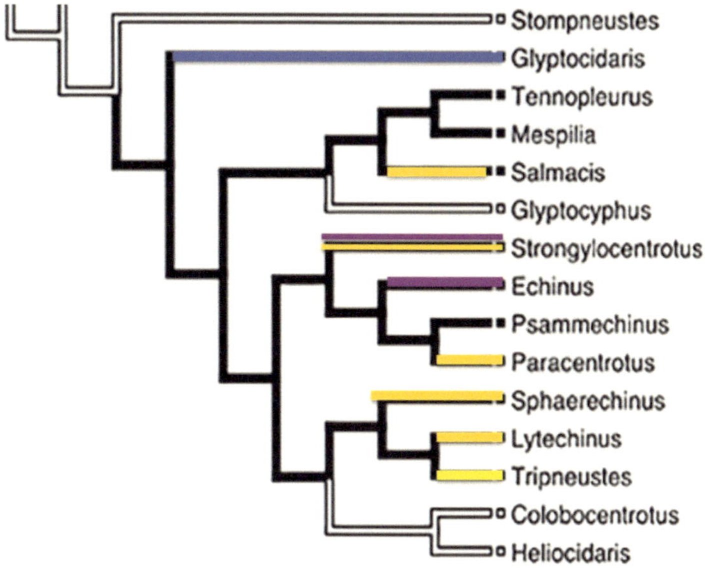

Figure 3. Covering behavior and evolutionary relationships. The uses of heaping by eleven urchin genera, mapped onto a phylogeny of urchin covering species courtesy of Ziegenhorn (2016). The colors represent different reasons for covering behavior.

shield themselves from light, and these urchins are closely related. *Strongylocentrotus* urchins, which use cover as a food source as well as protection from sunlight, are most closely related to *Echinus* urchins and *Paracentrotus* urchins, which use the reaction for the same reasons, as well as *Psammechinus* urchins, whose reason for crypsis is unknown.

In the *Salmacis-Mespilia-Temnopleurus* clade, relationships are less clear between genera as only *Salmacis* urchins' covering behavior has been explicitly studied (**Figure 3**). However, based on the phototaxic ability of both *Mespilia* and *Temnopleurus* urchins, it would be reasonable to conclude that these urchins also cover themselves for UV protection [6, 41]. *Glyptocidaris* is the only urchin genera in the phylogeny that heaps primarily for mechanical defense (**Figure 3**).

3.7. Covering by ocean depth

No relationship was found between the depth at which an urchin was found and its use for covering behavior (**Table 1**). For both 0–50 and 0–250 meter range of depths, UV protection was the most common reason for covering (**Table 1**). However, most urchins found in the range 0–100 meters had unknown reasons for heaping, and for the deeper ranging urchins (0–1000 and 0–1200 meters), light, mechanical damage, and predator protection were all reasons for covering (**Table 1**). These results suggest that there is no correlation between ocean depth of the urchin's environment and the function of their cryptic behavior.

Depth (meters)	Genera	Climate(s)	Climate(s)
0–50	*Paracentrotus*	Temperate	L
	Strongylocentrotus intermedius	Temperate/polar	L
	Toxopneustes	Tropical	M
0–100	*Psammechinus*	Polar	U
	Temnopeurus	Temperate	U
	Sphaerechinus	Temperate	L
	Strongylocentrotus purpuratus	Temperate	F
	Tripneuste	Tropical	L
	Pseudoboletia	Tropical	U
	Mespilia	Tropical/temperate	U
0–250	*Salmacis*	Temperate	L
	Glyptocidaris	Temperate	M
	Lytechinus	Tropical/temperate	L
0–1000	*Pseudechinus*	Temperate	P
	Genocidaris	Tropical/temperate	U
0–1200	*Echinus*	Temperate	F
	Strongylocentrotus drobachiensis	Temperate	L/M

NOTE: U stands for unknown, L stands for sunlight/UV protection, M stands for mechanical defense, and P stands for predator defense.

Table 1. Urchin covering by ocean depth: List of the 15 urchin genera and their use of covering behavior, organized by the depth ranges in which they are found.

3.8. Covering by climate

Though urchins who cover to protect themselves from light tend to be found in warmer oceans, this rule does not apply for all genera (**Figure 4a, b**). Some urchins within tropical regions cover for mechanical reasons (**Table 2**; **Figure 4b**). Additionally, some urchins use the behavior to shield themselves from light are also found in temperate and polar waters (**Table 2**). Based on the studies reviewed, no clear trend was found between urchin crypsis and ocean climate.

http://dx.doi.org/10.5772/intechopen.68469

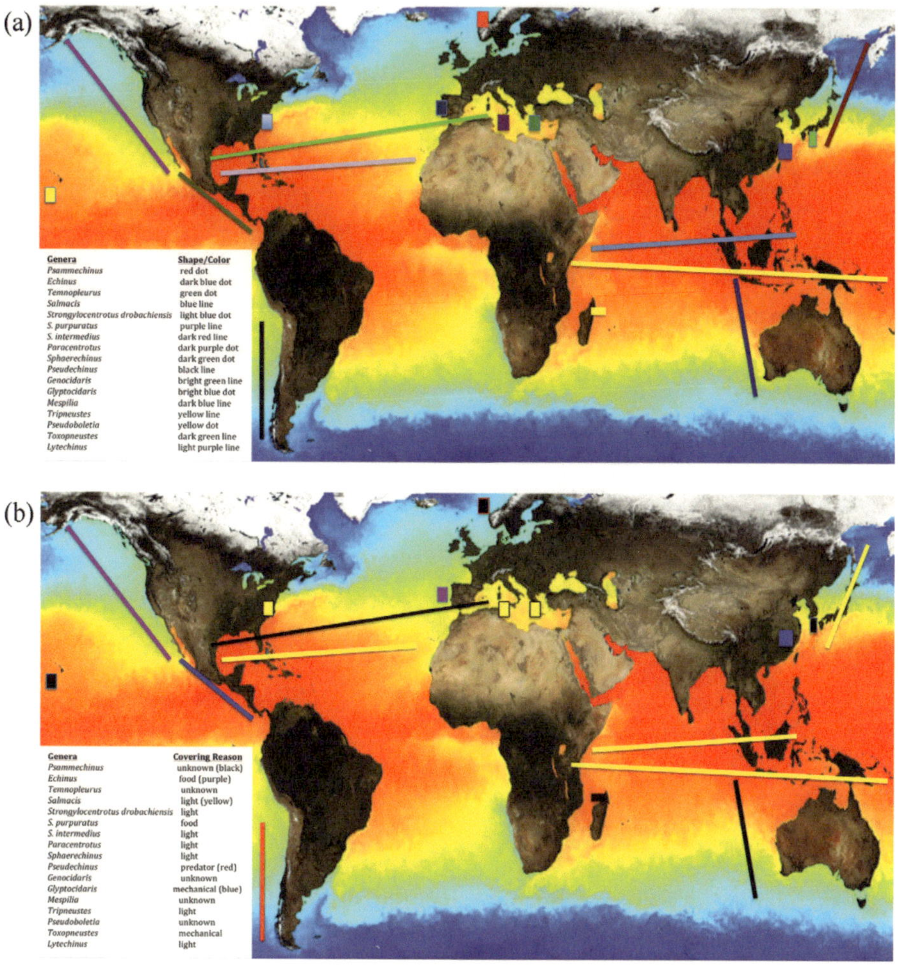

Figure 4. Map of urchins distribution and covering behavior. Distribution of the fifteen urchin genera considered, mapped onto an existing climate map (courtesy of NOAA). (a) Genera and shape/color, (b) genera and covering reason/ color. The covering reasons were abbreviated to unknown, light, mechanical, food, and predator.

Climate(s)	Genera	Depth (meters)	Covering Use
Tropical	*Tripneustes*	0–75	L
	Pseudoboletia	0–100	U
	Toxopneustes	2–50	M
Tropical/Temperate	*Genocidaris*	12–420	U
	Lytechinus	0–250	L
	Mespilia	0–60	U

Climate(s)	Genera	Depth (meters)	Covering Use
Temperate	*Echinus*	0–1200	F
	Temnopleurus	5–65	U
	Salmacis	0–180	L
	Strongylocentrotus drobachiensis	0–1150	L/M
	Paracentrotus	0–35	L
	Sphaerechinus	0–100	L
	Pseudechinus	0–820	P
	Glyptocidaris	10–150	M
	Strongylocentrotus purpuratus	0–65	F
Temperate/Polar	*Psammechinus*	0–100	Up
	Strongylocentrotus intermedius	0–35	L

NOTE: U stands for unknown, L stands for sunlight/UV shielding, M stands for mechanical protection, and P stands for predator defense.

Table 2. Urchin covering by ocean climate: List of the 15 urchin genera and their use of covering behavior, organized by the climate(s) in which they are found.

4. Discussion

Phylogenetic relationships are a better tool than environmental considerations for predicting why sea urchin species use cryptic behavior. Urchins that are found closer to the equator are more likely to cover to shield themselves from light, but *Toxopneustes, Pseudechinus,* and *Mespilia* urchins are exceptions and several genera including *Paracentrotus, Temnopleurus,* and *Strongylocentrotus* cover to protect themselves from light despite being further from the equator (**Figure 4a, b**). These exceptions decrease the power of climate as a predictive tool for determining the function of covering. This study also found no relationship between reason for the behavior and depth at which urchins were found (**Table 1**). In contrast, more closely related urchin species cover for similar reasons (**Figure 3**). This may suggest a genetic basis for urchin crypsis and its species-specific uses.

The covering reaction remains nebulous. For several genera, no studies have been performed to determine the function of the behavior, leaving gaps in the phylogeny as well as the climate and depth tables that might have impeded otherwise prominent patterns (**Figure 3**; **Table 1**; **Table 2**). It is also worth noting that for other genera that were assigned a most likely reason for covering, alternative hypotheses often were not tested. *Glyptocidaris* urchins, for example, were found to most likely heap for mechanical defense, and their place in the phylogeny suggests that this may be the ancestral reason for sea urchin crypsis (**Figure 3**). However, the effects of light intensity or the presence of predators have not been studied for *Glyptocidaris* and so cannot be conclusively ruled out as functions of their covering reaction. Of the urchin genera for which a most likely reason for covering was assigned, *Salmacis, Echinus, Sphaerechinus, Lytechinus, Toxopneustes,* and *Pseudechinus* all had at least one possible reason for the behavior that had not been explicitly tested. In other cases, multiple reasons for covering had convincing evidence, such as both food source and light protection for *Strongylocentrotus* urchins [7, 19].

It seems that heaping can perform several functions for a sea urchin, though within a genera there may be one primary reason for the behavior. *Tripneustes* and *Lytechinus* urchins primarily use the reaction to avoid bright light, but the importance of protection from wave surge has also been noted for both species [25, 33]. *Strongylocentrotus* urchins cover to shield themselves from light and use covering as a food source, but one study has also suggested the behavior might be used to prevent mechanical damage [17]. Because of this, it remains difficult to determine the reasons for a species cryptic behavior by any means apart from running experiments on the species in question. However, this study suggests that phylogenetic relationships provide a good framework for making predictions about the use of the covering reaction in other urchin species. For example, although the reason for the behavior is unknown for *Psammechinus* urchins, it is reasonable to predict that it also covers to defend itself from bright light because of its close relationship to *Paracentrotus* (**Figure 3**) and its noted phototaxic ability [40].

This study has important implications for the study of sea urchin heaping as a whole because of its novel suggestion that there might be a relationship between the various uses of covering and other aspects of urchin biology, be they genetic or environmental considerations. Understanding covering in sea urchins is particularly important because the ability is crucial to survival; one study has even shown that some sea urchins die if not provided with sufficient material with which to cover themselves [25].

Sea urchin ecology is important because of their role in coral reef systems, where they consume and control populations of algae that threaten corals [43], as well as in aquaculture, where they are part of the diet of humans in many countries [44]. Additionally, understanding the covering reaction itself, and how it may be determined by genetic and environmental factors, is important when considering it in relation to the multitude of cryptic behaviors that exist within the animal kingdom. Further studies of covering in urchin genera with currently unstudied covering behavior would be especially insightful for this study. Additional studies might also include ruling out other uses of the behavior for which one primary use has been considered, or studying the phylogenetic relationships of all urchin genera that cover.

Author details

Morgan A. Ziegenhorn

Address all correspondence to: maziegenhorn36@gmail.com

University of California, Berkeley, Berkeley, USA

References

[1] Agatsuma Y. Effect of the covering behavior of the juvenile sea urchin *Strongylocentrotus intermedius* on predation by the spider crab *Pugettia quadridens*. Fisheries Science. 2001;**67**(6):1181-1183

[2] Hughes TP, Reed DC, Boyle MJ. Herbivory on coral reefs: Community structure following mass mortalities of sea urchins. Journal of Experimental Marine Biology and Ecology. 1987 Jan 1;**113**(1):39-59

[3] Keegan BF, O'Connor BD, Könnecker GF. Littoral and benthic investigations on the west coast of Ireland: XX. *Echinoderm* aggregations. In. Proceedings of the Royal Irish Academy. Section B: Biological, Geological, and Chemical Science. Royal Irish Academy; 1985 Jan 1. pp. 91-99

[4] Verling E, Crook A, Barnes D. Covering behaviour in *Paracentrotus lividus*: Is light important?. Marine Biology. 2002 Feb 1;**140**(2):391-396

[5] Ziegenhorn MA. Best dressed test: A study of the covering behavior of the collector urchin *Tripneustes gratilla*. PloS One. 2016 Apr 13;**11**(4):e0153581

[6] Millott N. The photosensitivity of echinoids. Advances in Marine Biology. 1976 Dec 31;**13**:1-52

[7] Adams NL. UV radiation evokes negative phototaxis and covering behavior in the sea urchin *Strongylocentrotus droebachiensis*. Marine Ecology Progress Series. 2001 Apr 4;**213**:87-95

[8] Zhao C, Feng W, Tian X, Zhou H, Sun P, Chang Y. One generation of inbreeding does not affect covering behavior of the sea urchin *Strongylocentrotus intermedius*. Marine and Freshwater Behaviour and Physiology. 2013 Sep 1;**46**(5):345-350

[9] Dambach M. Covering reaction in sea urchins-new experiments and interpretations. Marine Biology. 1970 Jan 1;**6**(2):135

[10] Lawrence JM. Covering response in sea urchins. Nature. 1976 Aug;**262**:490-491

[11] Richner H, Milinski M. On the functional significance of masking behaviour in sea urchins an experiment with *Paracentrotus lividus*. Marine Ecology Progress Series. 2000 Oct 19;**205**:307-308

[12] Belleza DF, Abao RS, Taguba CA, Dy DT. Effects of UV-C on the masking behavior of the green urchin *Salmacis sphaeroides* (Linnaeus, 1758). The Philippine Scientist. 2012 Jan 1;**49**:34-43

[13] Wei J, Zhang L, Zhao C, Feng W, Sun P, Chang Y. Correlation analyses of covering and righting behaviors to fitness related traits of the sea urchin *Glyptocidaris crenularis* in different environmental conditions. Chinese Journal of Oceanology and Limnology. 2016 Nov;**34**:1183-1190

[14] Amato KR, Emel SL, Lindgren CA, Sullan KM, Wright PR, Gilbert JJ. Covering behavior of two co-occurring Jamaican sea urchins: Differences in the amount of covering and selection of covering material. Bulletin of Marine Science. 2008 Mar 1;**82**(2):255-261

[15] Sigg JE, Lloyd-Knight KM, Boal JG. UV radiation influences covering behaviour in the urchin *Lytechinus variegatus*. Journal of the Marine Biological Association of the United Kingdom. 2007 Oct 1;**87**(05):1257-1261

[16] Dayton PK, Rosenthal RJ, Mahen LC, Antezana T. Population structure and foraging biology of the predaceous Chilean asteroid *Meyenaster gelatinosus* and the escape biology of its prey. Marine Biology. 1977 Feb 1;**39**(4):361-370

[17] Dumont CP, Drolet D, Deschênes I, Himmelman JH. Multiple factors explain the covering behaviour in the green sea urchin, *Strongylocentrotus droebachiensis*. Animal Behaviour. 2007 Jun 30;**73**(6):979-986

[18] Ebert TA. Growth rates of the sea urchin *Strongylocentrotus purpuratus* related to food availability and spine abrasion. Ecology. 1968 Nov 1;**49**(6):1075-1091

[19] Crook AC. Individual variation in the covering behaviour of the shallow water sea urchin *Paracentrotus lividus*. Marine Ecology. 2003 Dec 1;**24**(4):275-287

[20] Berke SK, Miller M, Woodin SA. Modelling the energy–mortality trade-offs of invertebrate decorating behaviour. Evolutionary Ecology Research. 2006;**8**(8):1409-1425

[21] Verling E, Crook AC, Barnes DKA. The dynamics of covering behaviour in dominant echinoid populations from American and European west coasts. Marine Ecology. 2004; **25**(3):191-206

[22] Emlet RB. Developmental mode and species geographic range in regular sea urchins (*Echinodermata*: *Echinoidea*). Evolution. 1995 Jun;**1**:476-489

[23] Sharp DT, Gray IE. Studies on factors affecting the local distribution of two sea urchins, *Arbacia punctulata* and *Lytechinus variegatus*. Ecology. 1962 Apr 1;**43**(2):309-313

[24] Millott N. The covering reaction of sea-urchins I. A preliminary account of covering in the tropical echinoid *Lytechinus variegatus* (Lamarck), and its relation to light. Journal of Experimental Biology. 1972; **33**(3): 508-523

[25] Lees DC, Carter GA. The covering response to surge, sunlight, and ultraviolet light in *Lytechinus anamesus* (*Echinoidea*). Ecology. 1972 Nov 1;**53**(6):1127-1133

[26] Lewis JB. The biology of the tropical sea urchin *Tripneustes esculentus* Leske in Barbados, British West Indies. Canadian Journal of Zoology. 1958 Aug 1;**36**(4):607-621

[27] Kehas AJ, Theoharides KA, Gilbert JJ. Effect of sunlight intensity and albinism on the covering response of the Caribbean sea urchin *Tripneustes ventricosus*. Marine Biology. 2005 Apr 1;**146**(6):1111-1117

[28] Barnes D, Crook A. Quantifying behavioural determinants of the coastal European sea-urchin *Paracentrotus lividus*. Marine Biology. 2001 Jun 4;**138**(6):1205-1212

[29] Nahon S, Porras VA, Pruski AM, Charles F. Sensitivity to UV radiation in early life stages of the Mediterranean sea urchin *Sphaerechinus granularis* (Lamarck). Science of the Total Environment. 2009 Mar 1;**407**(6):1892-1900

[30] Unger B, Lott C. In situ studies on the aggregation behaviour of the sea urchin *Sphaerechinus granularis* Lam.(*Echinodermata*: *Echinoidea*). In Echinoderms through Time. Proceedings of the Eighth International Echinoderm Conference, Dijon, France; 1993 Sep. pp. 913-919

[31] Zhao C, Feng W, Tian X, Zhou H, Chang Y. Diel patterns of covering behavior by male and female *Strongylocentrotus intermedius*. Marine and Freshwater Behaviour and Physiology. 2013 Sep 1;**46**(5):337-343

[32] Zhao C, Zhou H, Tian X, Feng W, Chang Y. The effects of prolonged food deprivation on the covering behavior of the sea urchins *Glyptocidaris crenularis* and *Strongylocentrotus intermedius*. Marine and Freshwater Behaviour and Physiology. 2014 Jan 2;**47**(1):11-18

[33] James DW. Diet, movement, and covering behavior of the sea urchin *Toxopneustes roseus* in rhodolith beds in the Gulf of California, México. Marine Biology. 2000 Dec 12;**137**(5): 913-923

[34] Comely CA, Ansell AD. Population density and growth of *Echinus esculentus* L. on the Scottish west coast. Estuarine, Coastal and Shelf Science. 1988 Sep 1;**27**(3):311-334

[35] Forster GR. The ecology of *Echinus esculentus* L. Quantitative distribution and rate of feeding. Journal of the Marine Biological Association of the United Kingdom. 1959 Jun 1;**38**(02):361-367

[36] Douglas CA. Availability of drift materials and the covering response of the sea urchin *Strongylocentrotus purpuratus* (Stimpson). Pacific Science. 1976 Jan **1**

[37] Zhao C, Ji N, Zhang B, Sun P, Feng W, Wei J, Chang Y. Effects of covering behavior and exposure to a predatory crab *Charybdis japonica* on survival and HSP70 expression of juvenile sea urchins *Strongylocentrotus intermedius*. PloS One. 2014 May 16;**9**(5):e97840

[38] Ogden NB, Ogden JC, Abbott IA. Distribution, abundance and food of sea urchins on a leeward Hawaiian reef. Bulletin of Marine Science. 1989 Sep 1;**45**(2):539-549

[39] Pawson DL, Pawson DJ. Bathyal sea urchins of the Bahamas, with notes on covering behavior in deep sea echinoids (*Echinodermata*: *Echinoidea*). Deep Sea Research Part II: Topical Studies in Oceanography. 2013 Aug 31;**92**:207-213

[40] Millott N, Yoshida M. Reactions to shading in the sea urchin, *Psammechinus miliaris* (Gmelin). Nature. 1956 Dec 8;**178**(4545):1300

[41] Yoshida MA. Photosensitivity. In: Physiology of Echinodermata. New York. Interscience. p 435-464

[42] Yanagisawa Y. Preliminary observations on the so-called heaping behaviour in a sea urchin, *Hemicentrotus pulcherrimus* (A. Agassiz). Publications of the Seto Marine Biology Laboratory. 1972; **19**(6): 431-435

[43] Done TJ. Phase shifts in coral reef communities and their ecological significance. In. The Ecology of Mangrove and Related Ecosystems. The Netherlands: Springer; 1992. pp. 121-132

[44] McBride SC. Sea urchin aquaculture. In American Fisheries Society Symposium. American Fisheries Society; 2005 Vol. **46**. p. 179

Chapter 3

Effects of Environmental Factors on Reproduction of the Sea Urchin *Strongylocentrotus Intermedius*

Peter M. Zhadan, Marina A. Vaschenko and
Tatyana N. Almyashova

Additional information is available at the end of the chapter

http://dx.doi.org/10.5772/intechopen.69511

Abstract

The results of long-term studies (2003–2015) of the reproductive biology of the sea urchin *Strongylocentrotus intermedius* in wild populations located in the northwestern Sea of Japan along 400 km of the coast of the Primorye region of Russia and differing by the level of anthropogenic pressure are reported. Our analysis showed that since 1970–1980s, the shift in spawning season from autumn to early summer occurred in *S. intermedius* populations inhabiting anthropogenically polluted areas of Peter the Great Bay, resulting in the appearance of three types of populations that differ from each other in the proportions of individuals with early spawning (the end of May–June) and late spawning (September–early October). Our results indicate that neither photoperiod nor temperature may be considered as the primary external factors determining a shift in *S. intermedius* temporal patterns of gonad maturation and the timing of spawning and that phytoplankton concentration is the main factor for initiation of sea urchin spawning activity. We hypothesized that the shift in spawning season from autumn to early summer in *S. intermedius* populations inhabiting polluted areas can be explained by a phenotypic response of this species to environmental changes caused by chronic eutrophication.

Keywords: broadcast spawning, echinoderms, reproductive cycle, phytoplankton, temperature, anthropogenic pollution, eutrophication

1. Introduction

Reproduction of marine poikilothermic animals is a cyclic physiological process. Most marine invertebrate species release eggs and sperm into the water column where fertilization and development into a pelagic larval stage takes place and are called broadcast spawners. Broadcast spawners inhabiting the temperate climate zone mostly possess an annual reproductive cycle with the spawning period to the season when both abiotic and biotic factors (temperature, salinity, and phytoplankton) are favorable for offspring development and survival [1, 2]. The role of environmental variables in the regulation of the timing of the reproductive cycles and variations among and within populations of the same species is not fully understood. The temperature and photoperiod (either in combination or separately) are generally believed to be the most important environmental factors that control the succession of the reproductive cycle stages and spawning timing and promote the synchronization of the performance of these processes in different individuals within population [3, 4]. Mass synchronous spawning of the individuals of both sexes within a population is very important for reproductive success in broadcast spawners, that is determined by the short longevity of gametes and rapid gamete dilution associated with a decrease in fertilization efficiency [5–7]. It is known, however, that the reproductive cycle of some common species of sea urchins (*Paracentrotus lividus, Pseudechinus magellanicus, Strongylocentrotus droebachiensis, Strongylocentrotus intermedius*) is not synchronized within a population as well as between the populations [8–11]. Some individuals breed under conditions of rising photoperiod and increasing temperature whereas the others breed under opposite conditions. The reasons for such desynchronization of the reproductive cycle remain unclear.

It is well known that appearance of a large number of planktotrophic larvae of bottom invertebrates in the plankton coincides well with the occurrence of microalgae blooms [12–17]. As it was shown in laboratory experiments, the addition of microalgae and their extracts to sea water is able to induce the spawning of sea urchins and bivalve mollusks [18–20], and the combined addition of phytoplankton and sperm causes a synergistic effect on spawning activity [18, 20, 21]. Temperature [17, 22, 23], lunar phases [16, 21], and water salinity [24] can also serve as the stimuli that synchronize or trigger the spawning of bottom invertebrates. In addition, many researchers noted that a combination of several natural factors can contribute to spawning synchronization. Moreover, an analysis of the reproductive cycle of *P. lividus* [17] suggests that in sea urchins possessing fully mature gonads, spawning can begin even in the absence of external stimuli.

Sea urchins are widely distributed in marine coastal waters around the world. Many sea urchin species are both ecologically and economically important; besides, they are used as model organisms to study different aspects of reproductive biology of broadcast spawners both in the laboratory experiments and under field conditions. The sea urchin *S. intermedius* (A. Agassiz, 1863) inhabits hard substrates in the northern regions of the Asian Pacific coastal waters, from the Kamchatka Peninsula in the north to the Korean Peninsula in the south and from the Russian coast in the west to the Japanese Islands in the east [11, 25, 26].

In this paper, we give a review of the results of long-term studies (2003–2015) of the reproductive biology of the sea urchin *S. intermedius* in wild populations of this species located in the northwestern Sea of Japan along 400 km of the coast of the Primorye region of Russia and differing by the level of human activity. The main purposes of our work were: (1) to study the reproductive cycle and spawning schedule of *S. intermedius* populations in the areas with different level of anthropogenic pressure and (2) to clarify the relationships between natural environmental factors (phytoplankton, temperature, salinity, dissolved oxygen, moon cycle, tide level, and anthropogenic pressure) and the timing of spawning in sea urchin populations.

http://dx.doi.org/10.5772/intechopen.69511

2. Reproductive cycle and spawning schedule of *S. intermedius* in different populations

2.1. Study area, sea urchin sampling, and examination of gonadal state

Seasonal dynamics of gonadal state and the timing of spawning were studied in *S. intermedius* populations from 23 sites located in the northwestern Sea of Japan in the areas with different levels of anthropogenic pressure (**Figure 1** and **Table 1**). Ten sites in Amursky Bay and seven sites in Ussuriisky Bay, the secondary bays of Peter the Great Bay, are located at a different

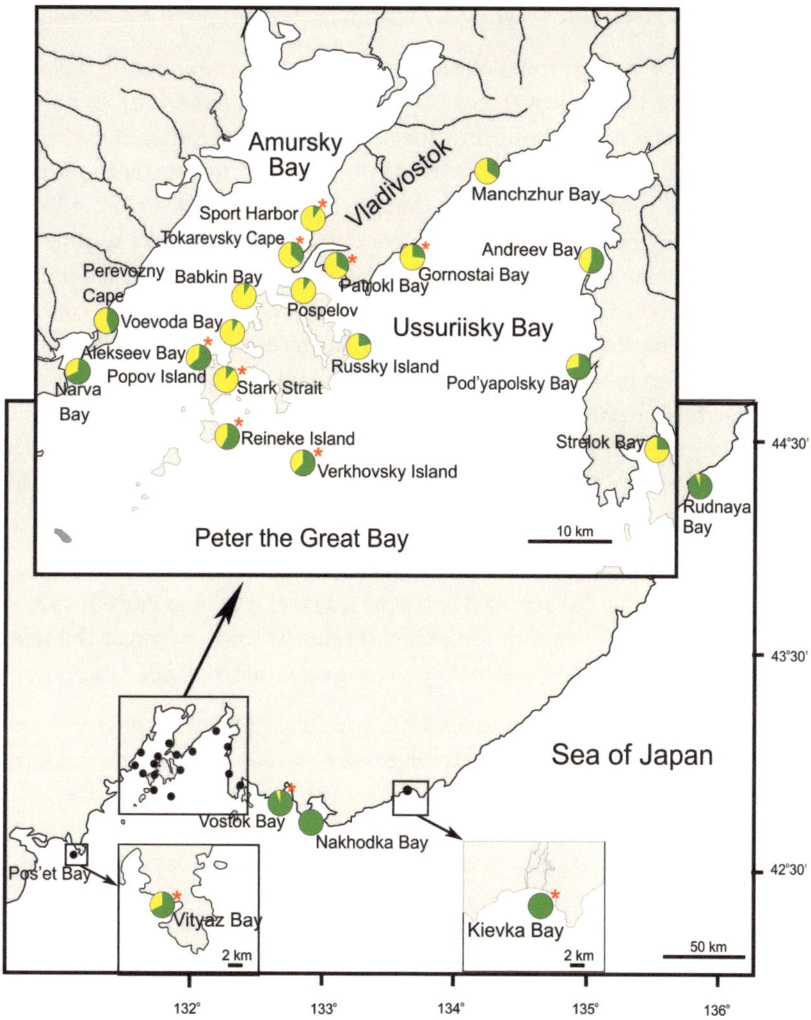

Figure 1. Map of the study area and spatial distribution of the populations of the sea urchin *Strongylocentrotus intermedius* differing in the proportions of individuals with different spawning schedules in the northwestern Sea of Japan. Light grey and dark grey parts of the circles correspond to the proportions of females with early (May–June) and late (September–early October) spawning. Asterisks denote the stations where the studies were conducted during more than one reproductive season.

distance from Vladivostok, the largest port in the Russian Far East of over 600,000 population, which is the main source of marine pollution in Peter the Great Bay. Numerous studies (reviewed in [27–30]) have been undertaken since the 1970s to determine the levels of anthropogenic pollutants in water, sediments, and biota of Peter the Great Bay and to assess the ecological consequences of pollution. Based on the results of these studies, we concluded that coastal waters adjacent to Vladivostok are chronically polluted by heavy metals of anthropogenic origin (Zn, Cu, Ni, and Pb) and organochlorine compounds such as dichlorodiphenyltrichloroethane (DDT) and polychlorinated biphenyls. The eastern coastal area of Amursky Bay where the sites Sport Harbor and Tokarevsky Cape are located and western coastal area of Ussuriisky Bay in the vicinity of a large municipal dump located on the Gornostai Bay coast (**Figure 1**) were considered the most polluted areas of Peter the Great Bay. The open part of Peter the Great Bay (off Russky, Popov, Reineke, and Verkhovsky Islands) was considered a relatively clean zone.

In the 1990s, there was an abrupt recession in industrial production in Primorye region that generated hope for the improvement of the ecological situation in Peter the Great Bay. Actually, according to chemical monitoring, the level of sediment contamination by oil, hydrocarbons, phenols, lead, and copper considerably decreased in the northern part of Amursky Bay [31]. However, our results of long-term monitoring of several sites located in the eastern coastal area of Amursky Bay showed that the contamination of bottom sediments by heavy metals and organochlorine pesticides in the late 1990s and in the early 2000s remained rather high [27]. Moreover, according to our data, during this period silting and contamination of bottoms in the island part of Peter the Great Bay increased in the vicinity of Popov Island (Amursky Bay) and Verkhovsky Islands (Ussuriisky Bay). This was probably related to the removal of thin sediment particles in the form of a suspension by currents from the coastal areas adjoining the city into the open island part of Peter the Great Bay. The ecological risk index SQG-Q, suggested by Long and MacDonald [32], was calculated for several surveyed areas in Amursky and Ussuriisky Bays [27]. It was found that sediments were potentially toxic at all surveyed sites.

Considering all above results, the Vostok Bay is a part of Peter the Great Bay whereas Kievka and Rudnaya Bays are located off Peter the Great Bay (**Figure 1**) were chosen as the reference sites in our studies of the reproductive cycle and spawning schedule in *S. intermedius* populations.

Sea urchins were sampled using scuba in 2003 and 2005–2010 on stony and stony-pebble grounds, at depth of 2–6 m once a month: annually at sites Sport Harbor, Gornostai Bay, Alekseev Bay, and Kievka Bay; from May to October at sites Rudnaya Bay, Patrokl Bay, Vostok Bay, Russky Island, Tokarevsky Cape, Verkhovsky Islands, Stark Straight, Reineke Island, Nakhodka Bay (Novitsky Bay site), and Pos'et Bay (Vityaz Bay site); from May to July at sites Perevozny Cape, Narva Bay, Manchzhur Bay, Andreev Bay, Pod'yapolsky Bay, and Strelok Bay (Abrek Bay site); in May and June, at several sites off Russky Island (Pospelov Beach, Babkin Bay, and Voevoda Bay), at the eastern side of Amursky Bay (**Figure 1** and **Table 1**). At 11 sites, the studies were conducted in sea urchin breeding seasons over more than 1 year (**Figure 1** and **Table 1**). The size of each sample was ≥50 individuals.

The gonads of 30 animals with test diameter of 50–75 mm were weighed to determine the gonad index (GI). The stage of gonad maturity was determined for each individual in accordance with

Station (N, E)	Year	Gonad index (mean ± SD)*	Period of spawning**		Proportion of females/males ready for spawning (%)	
			Early	Late	In the earlier period	In the later period
Rudnaya Bay (44°20′, 145°49′)	2009	24.0 ± 4.9	29.05–07.07	07.08–16.09	3/6	100/100
Kievka Bay(42°50′, 133°41′)	2007	18.6 ± 4.5	No spawning	28.08–10.10	0/0	98/100
	2008	20.0 ± 5.4	No spawning	21.08–25.09	0/0	100/100
	2009	19.5 ± 6.7	02.06–26.06	21.08–15.09	8/67	100/100
	2010	22.0 ± 4.1	No spawning	06.08–8.09	0/0	100/100
	2011	16.6 ± 1.9	No spawning	13.08–21.10	0/0	100/100
	2012	16.1 ± 3.4	No spawning	14.08–17.10	0/0	100/100
	2014	10.7 ± 6.7	No data	28.07–No data	0/0	100/100
	2015	14.1 ± 4.7	No data	13.08–No data	0/0	100/100
Nakhodka Bay						
Novitsky Bay (42°47′7″, 132°54′10″)	2010	24.1 ± 6.8	No spawning	19.07–27.09	0/0	100/100
Vostok Bay(42°53′41″, 132°43′59″)	2008	21.9 ± 5.7	19.06–27.07	12.08–21.09	18/29	95/95
	2009	17.8 ± 5.4	No spawning	02.08-22.10	0/0	90/100
Strelok Bay						
Abrek Bay (42°52′36″, 132°21′58″)	2010	14.1 ± 6.5	04.06–15.07	No data	74/100	No data
Ussuriisky Bay						
Pod'yapolsky Bay (43°0′6″, 132°17′56″)	2010	14.9 ± 6.1	04.06–18.07	No data	25/66	No data
Andreev Bay (43°6′24″, 132°19′20″)	2010	10.4 ± 4.1	04.06–18.07	No data	43/62	No data

Station (N, E)	Year	Gonad index (mean ± SD)*	Period of spawning**		Proportion of females/males ready for spawning (%)	
			Early	Late	In the earlier period	In the later period
Manchzhur Bay (43°14′3″, 132°13′51″)	2009	20.7 ± 4.7	11.06–29.07	No data	64/87	No data
Gornostai Bay(43°7′50″, 132°2′36″)	2009	19.4 ± 7.8	11.06–27.07	08.09–13.10	71/88	26/80
	2010	15.7 ± 5.0	15.06–22.07	21.09–20.10	72/73	22/75
Patrokl Bay (43°4′26″, 131°57′1″)	2009	16.5 ± 6.4	02.06–27.07	09.09–29.09	53/61	52/90
	2011	14.5 ± 4.9	25.05–08.07	05.08–06.10	76/75	27/31
Russky Island (43°1′25″, 131°56′8″)	2009	7.8 ± 3.6	25.05–13.06	14.09–26.10	77/59	18/56
Verkhovsky Islands (42°54′22″, 131°50′22″)	2009	14.9 ± 5.6	08.06–19.07	10.09–23.10	39/60	61/85
	2011	14.5 ± 4.7	13.05–14.07	8.09–25.10	40/41	60/59
Amursky Bay						
Sport Harbor(43°7′14″, 131°52′27″)	2003	23.3 ± 5.2	24.05–25.06	06.08–06.10	78/82	10/68
	2005	22.5 ± 4.8	31.03–04.06	02.08–06.09	82/80	10/62
	2006	27.0 ± 4.6	15.05–11.07	25.08–17.10	92/91	17/80
	2009	21.0 ± 4.8	28.05–07.07	06.09–22.10	92/98	43/95
	2010	23.8 ± 5.2	24.05–17.06	26.08–30.09	100/100	8/80
Tokarevsky Cape(43°5′5.10″, 131°50′45″)	2003	14.3 ± 3.2	24.05–25.06	06.08–06.10	54/68	52/79
	2013	13.3 ± 2.3	21.05–25.06	17.08–29.09	80/95	17/41
Pospelov Beach (43°03′30″, 131°52′34″)	2013	14.3 ± 3.3	16.05–21.06	No data	93/100	No data
Babkin Bay (43°03′08″, 131°47′27″)	2013	11.7 ± 5.7	16.05–21.06	No data	93/95	No data

http://dx.doi.org/10.5772/intechopen.69511

Station (N, E)	Year	Gonad index (mean ± SD)*	Period of spawning**		Proportion of females/males ready for spawning (%)	
			Early	Late	In the earlier period	In the later period
Voevoda Bay (43°00′39″, 131°46′14″)	2013	14.2 ± 2.3	16.05–20.06	No data	93/100	No data
Alekseev Bay(42°59′10″, 131°42′58″)	2003	7.2 ± 4.1	10.06–2.07	22.08–27.09	15/30	70/63
	2005	12.5 ± 4.7	17.04–10.06	01.09–12.10	44/60	51/45
	2006	13.6 ± 6.5	16.05–02.07	26.08–06.10	50/38	44/60
	2009	9.6 ± 3.9	27.05–16.06	11.09–13.10	38/47	54/44
	2010	11.1 ± 3.7	05.05–16.06	30.08–10.09	39/54	43/52
	2016	10.1 ± 4.9	22.05–10.06	17.08–12.09	58/70	33/45
Stark Straight(42°58′4″, 131°45′8″)	2009	14.4 ± 5.5	07.06–15.07	11.09–13.10	81/83	17/77
	2010	16.1 ± 6.4	05.05–18.06	10.09–22.10	96/75	8/73
Reineke Island(42°54′53″, 131°44′48″)	2003	12.5 ± 4.3	24.05–11.06	21.08–27.09	44/74	55/58
	2009	9.1 ± 3.8	08.06–19.07	12.08–23.10	38/56	50/70
Perevozny Cape (43°3′25″, 131°35′51″)	2010	27.4 ± 5.0	31.05–15.07	No data	55/97	No data
Narva Bay (42°57′49″, 131°30′19″)	2010	11.9 ± 5.3	31.05–15.07	No data	32/32	No data
Pos'et Bay						
Vityaz Bay(42°35′5″, 131°9′55″)	2007	15.3 ± 5.6	15.05–4.06	05.08–03.10	35/88	56/61
	2009	13.7 ± 6.3	25.05–6.07	No data	28/56	No data

*The gonad index (GI) data for sea urchin sample with maximum mean GI value are presented.

**The date of the spawning beginning corresponds to the time when the portion of spawned individuals was ≤10%. The date of the spawning ending corresponds to the time when the portion of spawned individuals was >90%.

Table 1. Reproductive characteristics of the populations of the sea urchin *Strongylocentrotus intermedius* in the northwestern Sea of Japan.

the slightly modified classification which was suggested for the sea urchin *S. intermedius* [33, 34]. Five stages of gonad maturity for both males and females were distinguished: (1) recovering, (2) growing, (3) premature, (4) mature, and (5) spent. Spawning was revealed by a significant increase in the number of spent individuals of both sexes and decrease in the GI.

2.2. Determination of temporal spawning patterns of different *S. intermedius* populations

An analysis of seasonal dynamics of gonadal condition showed the timing of spawning to be different in different *S. intermedius* populations of the northwestern Sea of Japan. We distinguished three types of sea urchin populations that differed from each other in the proportions of individuals with different spawning schedules (**Table 1** and **Figure 1**).

Sea urchin populations with pronounced late (autumn) spawning were referred to the first type. Sea urchin GIs in these populations were the highest during summer months and sharply decreased in September–October. The portion of females with mature gonads exhibiting egg release reached maximum in August to the beginning of September and sharply decreased in the second half of September–October, whereas the percentage of spent females in this period was 90–100% (**Table 1** and **Figure 2A**). Such populations are located from Vostok Bay north to Rudnaya Bay (**Figure 1**).

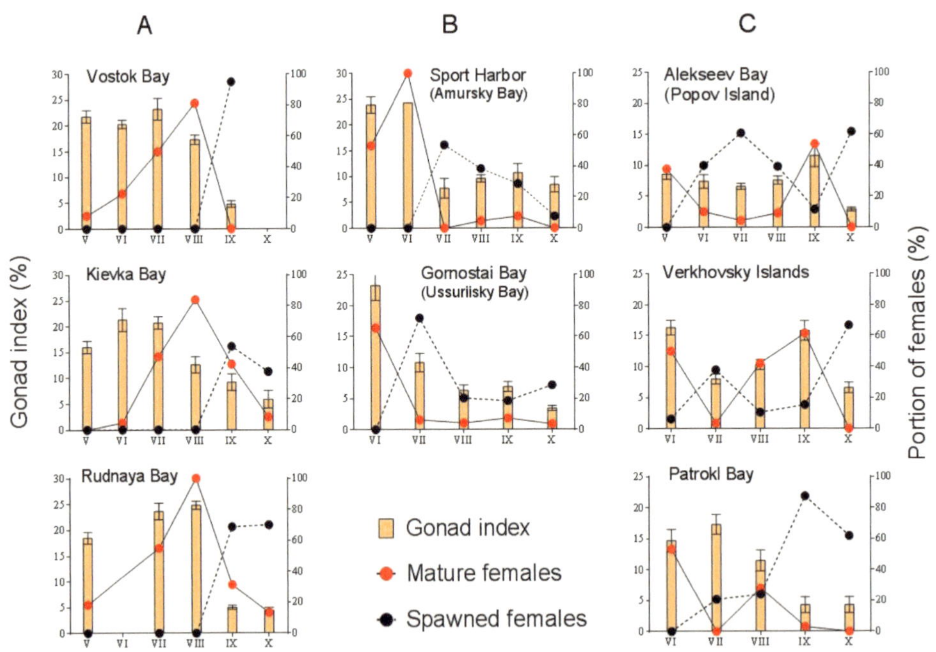

Figure 2. Seasonal dynamics of the gonad index (mean ± SE) and portions of mature and spawned females in populations of the sea urchin *Strongylocentrotus intermedius* with different spawning timing. (A) Populations of the first type (individuals with autumn spawning prevail); (B) populations of the second type (individuals with early summer spawning prevail); (C) populations of the third type (approximately equal proportions of individuals with early and late spawning).

Sea urchin populations with pronounced early spawning (early summer) were referred to the second type. Sea urchin GIs and the portion of mature females in these populations significantly decreased in June–July (**Figure 2B**). The portion of females that spawned in the end of May to the beginning of June varied from 71 to 100% in different years (**Table 1**). Such populations inhabit Sport Harbor and Gornostai Bay sites which are located in coastal waters of Amursky and Ussuriisky Bays adjacent to Vladivostok, as well as Stark Straight between Russky and Popov islands, Russky Island site at the western side of Ussuriisky Bay where heavily contaminated sediments dredged from Vladivostok harbor were dumped, and Abrek Bay located in Strelok Bay where Navy ships were based (**Figure 1**). In 2013, the second type populations were found at Pospelov Beach, Babkin Bay, and Voevoda Bay sites located in Amursky Bay off the Russky Island (**Figure 1** and **Table 1**).

The major part of *S. intermedius* populations belongs to the third type, which is characterized by approximately equal proportions of individuals with early and late spawning (**Figure 1**). In these populations, sea urchin GIs remained almost unchanged during summer and decreased in September–October, and the portion of mature females exhibited two picks, in May and August (**Figure 2C**). Such populations are located in the insular zone of Peter the Great Bay (Popov Island, Reineke Island, and Verkhovsky Islands) as well as at the sites near the western shore of Amursky Bay (Narva Bay and Perevozny Cape) and near the eastern shore of Ussuriisky Bay (Andreev and Pod'yapolsky Bays). Sea urchin populations from Manchzhur Bay, Patrokl Bay, and Tokarevsky Cape sites also can be referred to the third type; it should be noted, however, that at Tokarevsky Cape site in 2013, 80% of females and 95% of males with early spawning pattern were found (**Table 1**). At Vityaz Bay site located in Pos'et Bay, approximately 30% of females were ready for spawning in May–June (**Table 1**).

In most of *S. intermedius* populations studied, the individuals both with early and late spawning occurred, and their ratio was different in different locations (**Figure 1** and **Table 1**). Analysis of the dynamics of gonadal maturation and spawning in *S. intermedius* populations with different spawning schedules showed that the total proportions of females and males that spawned over the entire spawning season, including early and late periods, were 99 ± 9 and 126 ± 30%, respectively (mean ± SD, the differences were significant at $p < 0.0001$). This indicates that females participated in the spawning only ones, whereas significant proportion of males took part in both spawning periods. An exception was the total sample of sea urchins from Sport Harbor site collected in 2009 and comprised 92% females that had spawned in early summer and 43% females that had spawned in autumn period. However, the data on the timing of gonad maturation and spawning in females allowed us to determine the type of spawning (early-, late-, or double-peak) for *S. intermedius* populations [35].

The spawning schedule of *S. intermedius* did not depend on the relative weight of gonads. The largest GIs (>20%) were registered in sea urchins from sites Rudnaya Bay, Kievka Bay, Vostok Bay, and Sport Harbor (**Table 1**); however, late spawning was characteristic of the first three populations, whereas at the fourth site, early spawning was observed. In 2014, in Kievka Bay, sea urchins had relatively low GI of approximately 10% but retained the late spawning pattern. Significant variations in the GIs were found among *S. intermedius* populations with a double-peak (early and late) in spawning activity (**Table 1**).

Comparison of our results (**Table 1**) with literature data suggests that, at present, the spawning schedule of *S. intermedius* in the northwestern Sea of Japan differs from that described for this species in the end of 1960s, in 1970s, and the beginning of 1980s. The first studies of the reproductive cycle of *S. intermedius* in Russian waters of the Sea of Japan were performed in several secondary bays of Peter the Great Bay in 1970–1980s [36–39]. It was concluded that the spawning period of *S. intermedius* in Ussuriisky Bay and Vostok Bay usually falls on late summer–early autumn [38, 39]. In the 1970s, in *S. intermedius* populations inhabiting Pos'et Bay (at Pos'et Village and in Troitsa Bay) autumn spawning was also clearly pronounced; however, in 20–30% of sea urchin females the ripe eggs were also found in May [40]. In the end of 1970–1980s, approximately 5% of females with mature ovaries could be found in May in *S. intermedius* populations inhabiting Vityaz Bay (Pos'et Bay), whereas sea urchin mass spawning occurred from the end of August to October (Durkina V.B., NSCMB FEB RAS, personal communication).

According to our results, in 2007 and 2009 in Vityaz Bay, the portion of *S. intermedius* females which spawned in the early summer was on average 31.5%, and more than 50% of females spawned during late August to the beginning of October (**Table 1** and **Figure 1**). The reproductive cycle of *S. intermedius* inhabiting Amursky Bay has not been previously studied. However, in the mid-1980s to the beginning of 2000s, we observed an extremely low level of the gonad maturity in sea urchins from several sites in Amursky Bay in August, just before an anticipated *S. intermedius* mass spawning, and suggested another timing of spawning of this species [41, 42]. Indeed, our further studies revealed the peaks of early spawning (late May–June) in several *S. intermedius* populations from Amursky Bay (**Table 1** and **Figure 1**). The highest percentages of females participating in early spawning were found at sites Sport Harbor (88.8 ± 8.8%, 5 years mean ± SD, min = 78%, max = 100%) and Stark Straight (88.5 ± 10.6%, 2 years mean ± SD, min = 81%, max = 96%). This parameter tended to decrease along the western and eastern coasts of the bay, in the direction of its open part; however, several *S. intermedius* populations with more than 90% of females with the early spawning pattern were found in the vicinity of Russky Island (**Figure 1**).

In Ussuriisky Bay, the spawning schedule of *S. intermedius* has changed significantly in the last 40–50 years. In the late 1960s and early 1980s, autumn spawning was registered here [36, 38], while our studies revealed predominantly early spawning in most of *S. intermedius* populations. Thus, at Gornostai Bay site located close to municipal waste landfill, early spawning was registered in 71.8 ± 6.3% females (3 year mean ± SD, min = 65.5%, max = 78%). At Russky Island located close to dumping, 77% of females spawned during early summer (1 year observations). In Ussuriisky Bay, a tendency to decreasing percentages of females with early spawning along the western and eastern coasts of the bay was also observed, up to 30 and 40% at Pod'yapolsky Bay and Verkhovsky Islands sites, respectively (**Figure 1**). This tendency was disrupted in Strelok Bay adjacent to Ussuriisky Bay on the east, Abrek Bay is a part of Strelok Bay (area of Navy base), the portion of females with early spawning was 74%.

S. intermedius populations inhabiting Vostok Bay vicinal to the Strelok Bay and the bays located east and northeast of Vostok Bay exhibited predominantly late spawning (August–September). In May–June of 2009, the ripe eggs were found in the ovaries of only a small portion of sea urchin females in the samples from Vostok Bay (18%), Kievka Bay (8%), and Rudnaya Bay (3%) sites. It should be noted that the studies of the reproductive cycle of *S. intermedius* conducted in 1971–1975 in Vostok Bay revealed only late spawning of *S. intermedius* [37].

Thus, our results showed that the reproductive cycle of *S. intermedius* in the study area is desynchronized both within one population (the individuals spawn in different seasons, in early summer and autumn) and between the populations (there are three types of populations differing in the proportions of individuals with early or late spawning).

2.3. Hypotheses to explain regional differences in sea urchin spawning schedule

The studies of the reproductive cycle of *S. intermedius* from different population in the Hokkaido area along the Sea of Japan, Sea of Okhotsk, and Eastern Pacific sides conducted over the past 50 years have also revealed different spawning schedules in this species (see for review [11]). Three patterns of *S. intermedius* reproductive cycle were distinguished, as follows: (1) "Sea of Japan cycle" with autumn spawning peak (September–October); (2) "Sea of Okhotsk-Eastern Pacific cycle" with extended spawning period (June–October), and (3) a cycle with two pronounced spawning peaks, in spring (April–May) and in autumn (August–October) that was characteristic of sea urchins inhabiting the northern part of Tsugaru Strait and Funka Bay (southern coast of Hokkaido) [11]. It is important to note, however, that Fuji [33] studied the reproductive cycle of *S. intermedius* in the eastern Tsugaru Strait and Funka Bay in 1950s and revealed only autumn spawning peak (September–October).

There is evidence that the populations of other sea urchin species, in some areas of their geographic range, can spawn in different seasons [8–10]. In most detail, the reproductive biology has been studied in the sea urchin *P. lividus* distributed along the western coast of the Atlantic, from Ireland to the southern extremity of Morocco and in the Mediterranean. The data from 52 publications were involved into a statistical analysis to reveal spatial and temporal patterns of the reproductive processes of this species over the entire geographical range [43]. It was shown, that Atlantic populations of *P. lividus* spawned once a year, in spring–early summer. Mediterranean populations of this species, according to most authors, spawned twice (in spring and autumn), although some authors reported either a single spring spawning or multiple summer spawning peaks [44, 45].

The reasons for such desynchronization of the reproductive cycle of the same sea urchin species remain unclear. There are several hypotheses that have been proposed to explain regional differences in spawning schedule of sea urchins.

1. Hypothesis of the latitudinal gradient is based on the recognition of temperature and photoperiod as the main environmental factors regulating reproductive cycles of marine invertebrates and states that in temperate waters, the specimens spawn during a short period of the year while under tropical conditions, the spawning period extends and some species can spawn throughout the year (see for review [46]). However, this hypothesis cannot explain, why *P. lividus* populations located along the western coast of the Atlantic, from Ireland to the southern extremity of Morocco, that is, for approximately 2000 km from

south to north, are characterized by a single spawning [43], whereas the populations of the same species inhabiting the Mediterranean and located at the same latitude have several spawning peaks that mostly fall on spring and autumn [44, 45]. This hypothesis also cannot explain different spawning schedules of the sea urchin *S. intermedius* in Peter the Great Bay (Sea of Japan) because all the examined populations of this species were located at approximately the same latitude (**Figure 1** and **Table 1**). The populations of this species around Hokkaido are also located at approximately the same latitude but have different spawning seasons (see for review [11]).

2. Hypothesis of genetic determination of temporal patterns of gonad maturation and timing of spawning of *S. intermedius* is based on the results of the experiments on transplantation of the offspring obtained from sea urchins with different spawning schedules [47, 48]. In these experiments, the offspring of sea urchins inhabiting the Sea of Japan side of Hokkaido was transplanted into the region of the eastern Pacific coast and *vice versa*, the offspring of sea urchins from the oceanic region into the Sea of Japan. The results showed that in both cases sea urchins retained the parental pattern of the reproductive cycle. It was suggested that these two populations are isolated. However, from this point of view, it is difficult to explain the changes in spawning schedules of some *S. intermedius* populations happened during the recent 50–60 years. For example, in the 1950s the spawning in sea urchin populations of southern Hokkaido occurred from September to November [33], whereas studies of 1980–1990s revealed double spawning (spring and autumn) in this area [11]. This disagreement was explained by changes in the warm Tsushima Current and Oyashio Current, which provided prerequisites for transfer of sea urchin larvae from other regions of the Pacific. In the northwestern Sea of Japan, in the late 1960s and early 1980s autumn spawning was also registered in *S. intermedius* from Ussuriisky Bay [38], whereas in the 2000s we revealed double spawning in this species [35]; moreover, in the population at Gornostai Bay site, most sea urchins spawned during early summer (**Table 1**). From point of view of the hypothesis mentioned above [47, 48], it should be concluded that genetically isolated populations of *S. intermedius* exist in Peter the Great Bay at distances of a few tens kilometers from each other. However, taking into account active hydrodynamics of this area owing to monsoon climate, such a conclusion appears unlikely. Summer winds of mostly southern directions and northerly autumn winds initiate pronounced wind currents mixing the water and transferring planktonic invertebrate larvae from the open part of Peter the Great Bay to its inner areas and *vice versa*. Also, no information is available about any changes in hydrological regime of the bay for the recent 50 years, which could benefit the transfer of *S. intermedius* larvae from remote areas, like, for example, the Sea of Okhotsk.

Moreover, the study of the genetic structure of 10 *S. intermedius* populations in the northwestern Sea of Japan that differ in the proportion of individuals with different spawning seasonality was conducted using seven allozyme loci as genetic markers [49]. No significant genetic differences between specimens of *S. intermedius* with different timing of spawning were revealed (genetic similarity was 0.988–0.991). A similar conclusion was drawn from the study of population genetic structure of 8 *S. intermedius* populations with different spawning schedule based on the analysis of 12 polymorphic loci of microsatellite DNA [50]. The results of these studies led us to suggest that the shift of spawning period of

S. intermedius in Peter the Great Bay from autumn to early summer is a phenotypic response of sea urchin populations to changes in environmental conditions.

3. Hypothesis of phenotypic response of *S. intermedius* populations to changing environmental conditions due to chronic anthropogenic pollution is based on our observations that: (1) early spawning is the most characteristic of sea urchin populations located close to sources of pollution and (2) during the recent 50–60 years, the shift in spawning season from autumn to early summer happened in some sea urchin populations inhabiting chronically polluted environments in Peter the Great Bay, whereas in relatively clean areas, sea urchin populations retained autumn spawning [49].

2.4. Marine pollution, eutrophication, and sea urchin reproductive cycle

Sea urchins are characterized by high level of phenotypic plasticity; their morphological and physiological characteristics (growth rate, maximum sizes of body and gonads, and morphology of the body) are prone to changes during the adaptation to particular environmental conditions [51]. Boudouresque and Verlaque [52] explained different spawning schedules in different populations of the sea urchin *P. lividus* in terms of an adaptive response of this species aimed to increase the chances of the offspring for survival. The individuals of *P. lividus* are supposed to synchronize their spawning with conditions favorable for food supply for the larvae (availability of phytoplankton) and preventing their evacuation with currents into the open ocean. An analysis of genetic variability using fragments of mitochondrial and nuclear DNA as markers in two cohorts of *P. lividus* recruits enriching the populations of this species in the Mediterranean after spring and autumn spawning peaks did not reveal any significant differences between the cohorts [53]. The results of our study, which showed the absence of significant genetic differences between the individuals of *S. intermedius* from Peter the Great Bay that spawn in early summer and in autumn [49, 50], are consistent with these data and greatly support the hypothesis that different spawning timing in different populations of the same sea urchin species can be result of their adaptation to environmental changes caused by chronic anthropogenic pollution of marine environment.

Significant positive correlation was found between the portions of females with early spawning in *S. intermedius* populations from several localities in Peter the Great Bay and potential toxicity of surface bottom sediments from these localities assessed by the calculation of the ecological risk index (**Figure 3**) that supports this hypothesis.

It is important to note that marine environment pollution is associated with eutrophication of waters; thus, the effects of anthropogenic pollution on the reproductive cycle of the sea urchin are likely to be indirect and conditioned by great concentrations of phytoplankton and its metabolites in seawater. Based on the analysis of the structural and quantitative characteristics of phytoplankton of Peter the Great Bay during the period from 1996 to 2009, the waters adjacent to Vladivostok in the Amursky Bay were characterized as extremely eutrophic (>3×10^6 cells/l) whereas the open waters of Amursky Bay and Ussuriisky Bay as well as Vostok Bay were attributed to the eutrophic type (from 3×10^4 to 3×10^6 cells/l) [54]. The changes in phytoplankton composition and biomass in Amursky Bay are attributed, to a significant degree, to anthropogenic eutrophication [55].

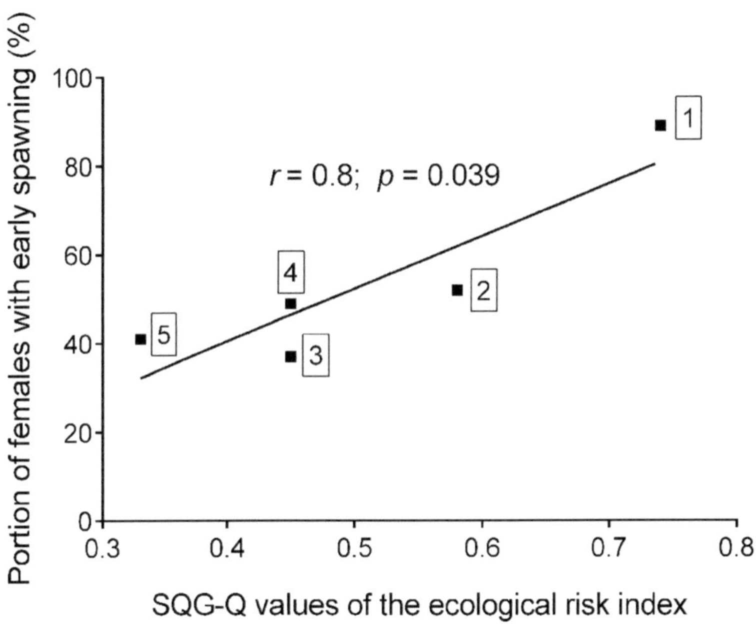

Figure 3. Correlation between the portions of females with early spawning in populations of *Strongylocentrotus intermedius* inhabiting several localities in Peter the Great Bay and potential toxicity of surface bottom sediments from these localities assessed by the calculation of the ecological risk index SQG-Q. 1—Sport Harbor, 2—Tokarevsky Cape, 3—Alekseev Bay, 4—Verkhovsky Islands, 5—Reineke Island. The SQG-Q values are given according to Ref. [27].

To address whether the timing of spawning in *S. intermedius* populations is associated with the level of anthropogenic pressure in the study area, we examined in more details the reproductive cycle of this species in Kievka Bay. This bay is located in sparsely populated area adjoining the State Reserve of Laso, one of the largest forest reserve in Russian Far East; therefore, there is no any significant anthropogenic impact on the marine environment in the chosen area. The wind-induced upwelling associated with the monsoon has a pronounced effect on the hydrological regime and enhances the primary productivity in this bay, especially during the period from April to October [56–59].

In the course of our research, we have faced an unexpected phenomenon. In 2008, more than 90% of sea urchin specimens seemed to have not completed their reproductive cycle by spawning [59]. To the best of our knowledge, this was the first case when the spawning failure in wild sea urchin population was revealed. Below we give a brief review of this phenomenon and discuss potential environmental causes.

3. Spawning failure in *S. intermedius* populations

Sea urchins were sampled in Kievka Bay from 2 to 6 m depths monthly from April 2008 to April 2011, that is, across three reproductive cycles. Methods for GI determination and gonad histology are described in Ref. [59]. In addition to five stages of gonad maturity (1) recovering, (2) growing, (3) premature, (4) mature, and (5) spent, we distinguished one more stage for unspawned females, in which the gonadal cleaning process was highly prolonged. Therefore, we referred to this period of the reproductive cycle as the stage of prolonged oocyte resorption (6) [59]. In unspawned males, the gonad acini contained a large number of spermatozoa across two reproductive cycles, therefore, we referred to this period as the mature stage (4). The percentage of ready for spawning individuals that released gametes through gonopores during sampling or after dissection was calculated and referred to a group of sea urchins ready for spawning, independently whether they released normal or degenerating gametes.

3.1. Peculiarities of gonadal temporal dynamics in spawned and unspawned *S. intermedius*

Examination of *S. intermedius* gonadal development throughout three reproductive cycles showed that in all the years, the GI of *S. intermedius* reached its maximum in the period from April to July (**Figure 4**) which corresponds to growing (2) and premature (3) stages of gonadal development. In August, the gonads of both sexes were at the premature (3) and mature (4) stages of the reproductive cycle (see **Figure 3** in Ref. [59] for temporal dynamics of the percentages of the reproductive cycle stages). However, despite the seasonal ripening of gonads

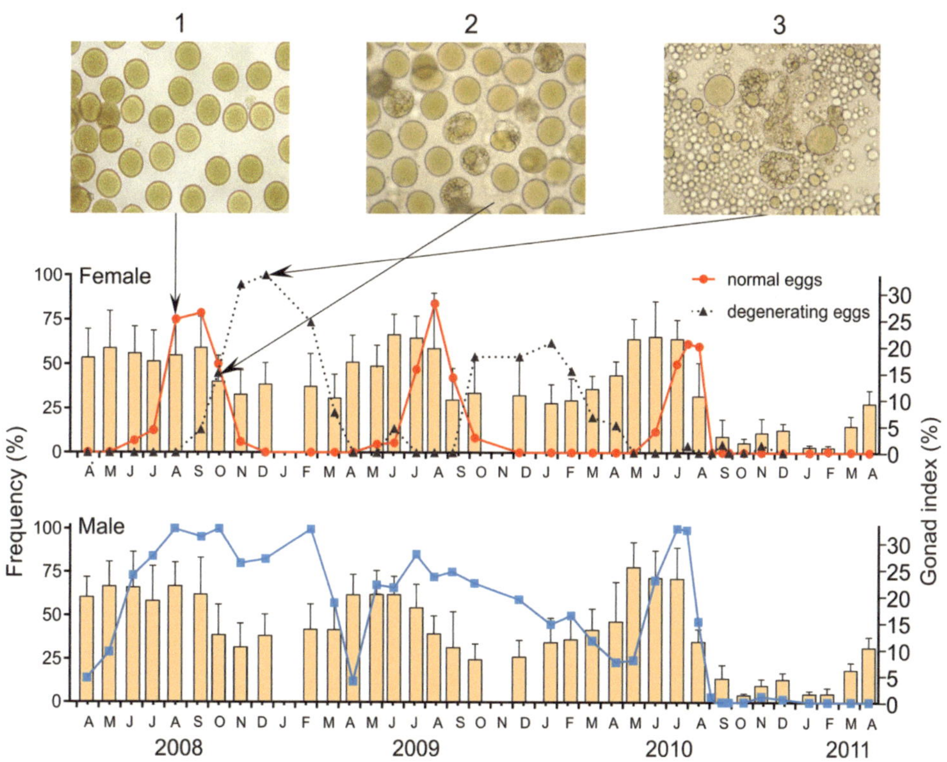

Figure 4. Temporal dynamics of the gonad index (bars, mean ±SE) and the percentages of *Strongylocentrotus intermedius* females (lines with circles and triangles) and males (lines with squares) ready for spawning from Kievka Bay (northwestern Sea of Japan) in 2008–2011. Photograph 1 shows normal eggs, photograph 2 and 3 show degenerating eggs at different stages of their fragmentation. The data shown are extracted from **Figures 3** and **5** and the graphical abstract in Ref. [59].

by all individuals in *S. intermedius* population, the dynamics of the GIs and percentages of females and males ready for spawning as well as gonad histology throughout April 2008 to April 2011 indicate that spawning may not occur for a most individuals in a population in some years. In two of three reproductive seasons, a nearly complete absence (95% unspawned females in 2008) or a partial absence (53% unspawned females in 2009) of spawning was observed. The males also did not spawn completely. In September, the destruction of undischarged eggs had begun in unspawned females. Undischarged eggs disintegrate into numerous spherical fragments (**Figure 4**). The stage of prolonged oocyte resorption (6) as well as the cleaning process associated with the resorption of numerous undischarged sperm by testicular nutritive phagocytes lasted for approximately 7 months [59].

The seasonal dynamics of the percentages of sea urchins ready for spawning generally corresponded to the data obtained from the histological analysis and reflected well the level of gonadal maturity [59]. In August 2008, approximately 80% of females released normal eggs; in September, the percentage of females ready for spawning reached 91% but 13% of females released a mixture of degenerating and normal eggs (**Figure 4**). In December, the percentage of females ready for spawning reached 100%; however, only degenerating eggs and egg fragments were released. The females which released degenerating eggs were observed through March 2009 (**Figure 4**). In 2009, the dynamics of females ready for spawning was similar to that in 2008; however, the percentage of females releasing degenerating eggs during autumn–winter seasons was two times lower.

Judging from the 3 month shift in the line reflecting the percentage of females which released fragmented eggs compared with that reflecting the percentage of females which released normal eggs (**Figure 4**) and from significant positive correlation between these dynamics ($r = 0.966$, $p = 0.0073$), it may be proposed that a life-time of mature eggs in sea urchin ovary lasts approximately 3 months.

The males ready for spawning were found in all sea urchin samples taken in 2008 and 2009 (**Figure 4**). During the period from August 2008 to March 2009, the portion of males ready for spawning was approximately 100%, and during the period from July 2009 to December 2010, it was from 87 to 57%. In July–August 2010, the percentage of males ready for spawning increased up to 100%.

In 2010, the GIs decreased stepwise from August to October up to zero in both sexes, and the percentage of sea urchins ready for spawning dropped to zero in September (**Figure 4**). It indicates that spawning in *S. intermedius* population was synchronous in females and males and was completed by October. The 3-year minimum for the GI was observed during the period from October 2010 to February 2011 (**Figure 4**), when the GIs were 5.8 and 4.5 times lower than the corresponding GI values recorded in 2008 and 2009, respectively.

The data on the size frequency distribution of female reproductive cells (oocytes and eggs) provided valuable information regarding the dynamics of oocyte development over three reproductive cycles, in relation to the presence or absence of spawning in *S. intermedius* population (**Figure 5**). Generally, the changes in the oocyte size frequency distribution between March and August were similar for the 3 years, reflecting a single cohort of growing oocytes

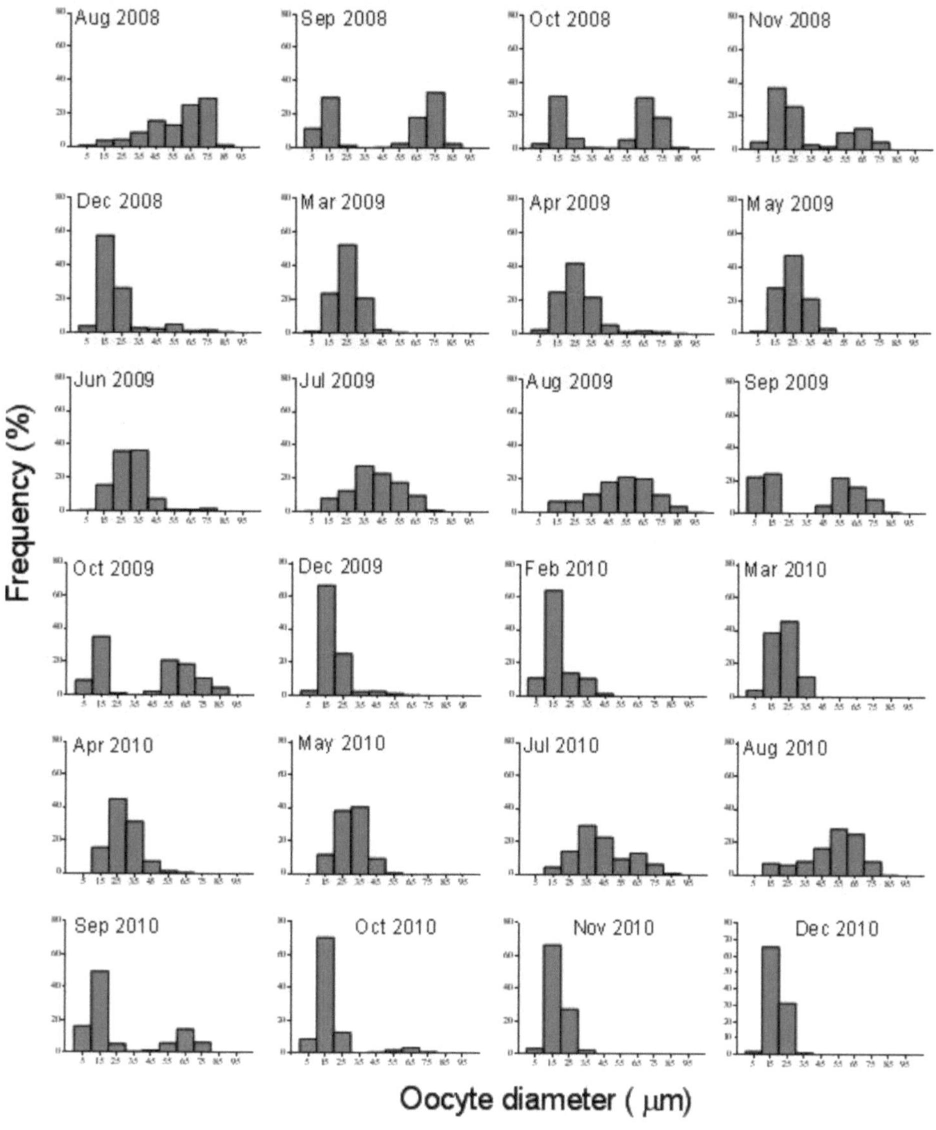

Figure 5. The size frequency distribution of the reproductive cells in *Strongylocentrotus intermedius* females sampled in Kievka Bay (northwestern Sea of Japan) in 2008–2010.

which was present in the ovaries, and the oocyte size distribution was approximately unimodal. However, some peculiarities were observed in the period from August to December. It is important to note that in all the 3 years, a fraction of the smallest oocytes that just began to grow (a diameter of approximately 10 μm) was absent in June–July and appeared in September, independent of whether spawning in *S. intermedius* population took place or not. In September and October of 2008 and 2009, the oocyte size distribution was bimodal, reflecting approximately equal shares of small growing oocytes with a diameter <30 μm and large vitellogenous oocytes with a diameter >50 μm and undischarged eggs (**Figure 5**). In November, when no morphologically normal eggs were present in the gonads (fragmented eggs were not taken into account), a fraction of large oocytes decreased sharply, and in December, it was negligible. In September and October of 2010, after complete spawning, spent ovaries presented 70–80% of growing oocytes and less than 20% of residual eggs which disappeared by November (**Figure 5**).

Thus, our study revealed the unique characteristics of the reproductive cycle of *S. intermedius* individuals displaying spawning failure. First, in the case of normal spawning, a more than 10-fold decrease in the GI occurred in September–October whereas in the case of absence of complete spawning, the decrease in the GI was about half of its maximum value, and these intermediate GI values remained stable (no significant differences between subsequent samples) until May of the next year. Second, the cleaning process associated with the resorption of numerous undischarged eggs and spermatozoa was unusually long, lasting for approximately 7 months.

It is well known that after spawning, the acini of sea urchin gonad still contain small quantity of undischarged gametes. These gametes subsequently undergo resorption, which involves nutritive phagocytes [33, 60–64]. Our data indicate that this mechanism is triggered regardless of whether spawning occurs. The ovaries and testes of unspawned sea urchins contain a tremendous quantity of undischarged eggs and sperm which have to be subjected to phagocytosis. Due to the differences in size of female and male gametes, the patterns of their phagocytosis are also different. Undischarged spermatozoa are phagocytosed by testicular nutritive phagocytes and become a part of their heterophagosomes, that is, undergo intracellular digestion [61–64]. Undischarged eggs are too large to undergo phagocytosis without being disintegrated. The large-scale fragmentation of undischarged eggs which we revealed in the ovaries of unspawned *S. intermedius* [59] corroborate the suggestion of Masuda and Dan [65] that sea urchin residual eggs are destructed into spherical fragments, probably through autophagy, which subsequently are phagocytosed by nutritive phagocytes and digested.

The resorption of relict gametes in the sea urchin gonad normally lasts for 2–3 months [64]. The much greater duration (over 7 months) of the cleaning process revealed in the ovaries of unspawned *S. intermedius* [59] is in all likelihood due to the much larger volume of the sex cells that were subject to resorption in seasons of 2008–2009. Development of a new generation of oocytes in *S. intermedius* took place from December to August (**Figure 5**). Thus, in the absence of spawning, the growth and differentiation of the new generation of oocytes occur simultaneously with the cleaning process.

3.2. Occurrence of spawning failure in *S. intermedius* populations along the coastline of the Primorye region and dynamics of environmental variables

To reveal the spatial (geographic) distribution of cases of spawning failure in *S. intermedius* populations in the northwestern sea of Japan, we carried out the following studies: (1) a one-time survey to examine the gonads of sea urchins sampled from Vostok Bay (42°53′ N, 132°43′ E) in November 2008 (in the post-spawning period of *S. intermedius* reproductive cycle); (2)

monthly analysis of sea urchin gonads during the period from May to November 2009 in Vostok Bay and Rudnaya Bay (44°20′ N, 145°49′ E); (3) a one-time analysis of the sea urchin gonads sampled in November 2009 in the Bays of Vrangel (42°45′ N, 133°3′ E), Shepalova (42°41′ N, 133°4′ E), Sokolovskaya (42°52′ N, 133°53′ E), and Kit (43°4′ N, 134°11′ E) (**Figure 6**).

Our interest to Vostok Bay is explained by the fact that (1) in this bay, the reproductive cycle of *S. intermedius* has been examined earlier for 5 years, in 1971–1975, and spawning failure has not been observed [37]; (2) according to satellite data (http://www.satellite.dvo.ru), the chlorophyll *a* (Chl *a*) concentration in Vostok Bay significantly exceeded that in Kievka Bay.

In sea urchin sample taken from Vostok Bay in November 2008, only 1 of 14 females (7%) was unspawned (**Figure 6**) and had a large gonad with degenerating eggs. Seasonal analysis of gonadal state showed that in 2009, sea urchins in Vostok Bay and Rudnaya Bay (extreme south and north sites) as well as in Kievka Bay (centrally positioned site) became mature almost simultaneously. In the first half of August, more than 90% of ripe sea urchins were found in all these populations, which allows to assume synchronous maturation in *S. intermedius* populations at other sites (the Bays of Vrangel, Shepalova, Sokolovskaya, and Kit). However, the percentage of unspawned females varied from 0% in Vostok and Sokolovskaya Bays to 68% in Kit Bay (**Figure 6**).

To reveal regional differences in the temperature of the water surface and phytoplankton production (determined as the Chl *a* concentration) in the surveyed area, satellite monitoring

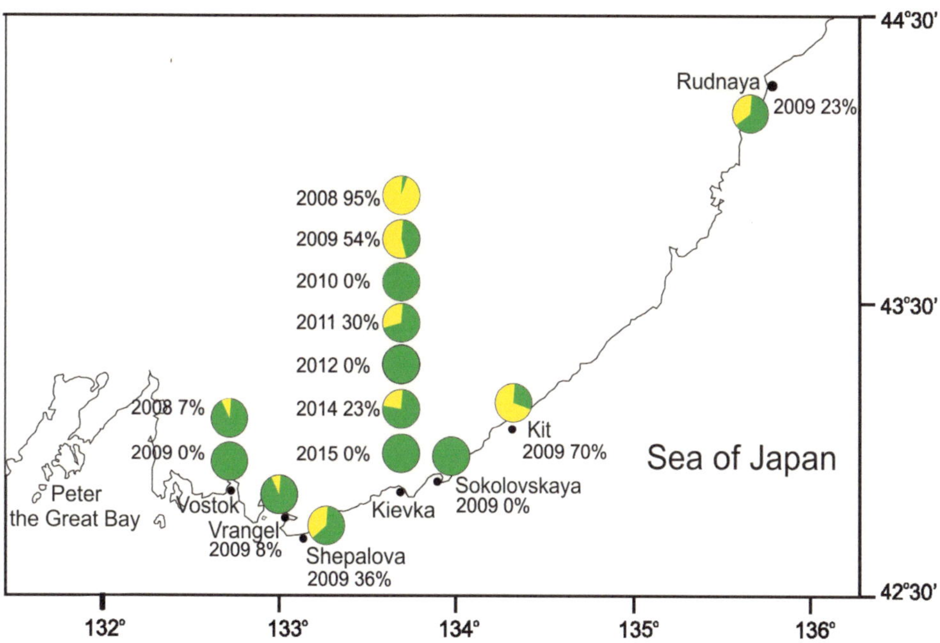

Figure 6. Occurrence of spawning failure in *Strongylocentrotus intermedius* populations along the coastline of the Primorye region (northwestern Sea of Japan). Yellow parts of the circles denote the percentages of unspawned females. The data shown are taken from Ref. [59], with addition of new data.

data over the period of 2008–2009 were used (see Ref. [59] for details). From our point of view, the comparison of temperature and Chl *a* dynamics in Kievka Bay and Vostok Bay is of most interest due to a large difference in the percentages of unspawned females in these bays.

The temperature profiles in Kievka and Vostok Bays during August and the beginning of October 2008 were well synchronized and strongly correlated (Pearson $r = 0.81$, $p < 0.0001$, $\alpha = 0.05$) (**Figure 7A**). An abrupt temperature decrease caused by upwelling, which is typically observed in this season [59], occurred on September 28 at both stations simultaneously. At the same time, the concentrations of Chl *a* greatly differed among these two sites: the maximum and mean values in Vostok Bay were 5.5 and 3.5 times higher, respectively, than those in Kievka Bay (**Figure 7B**).

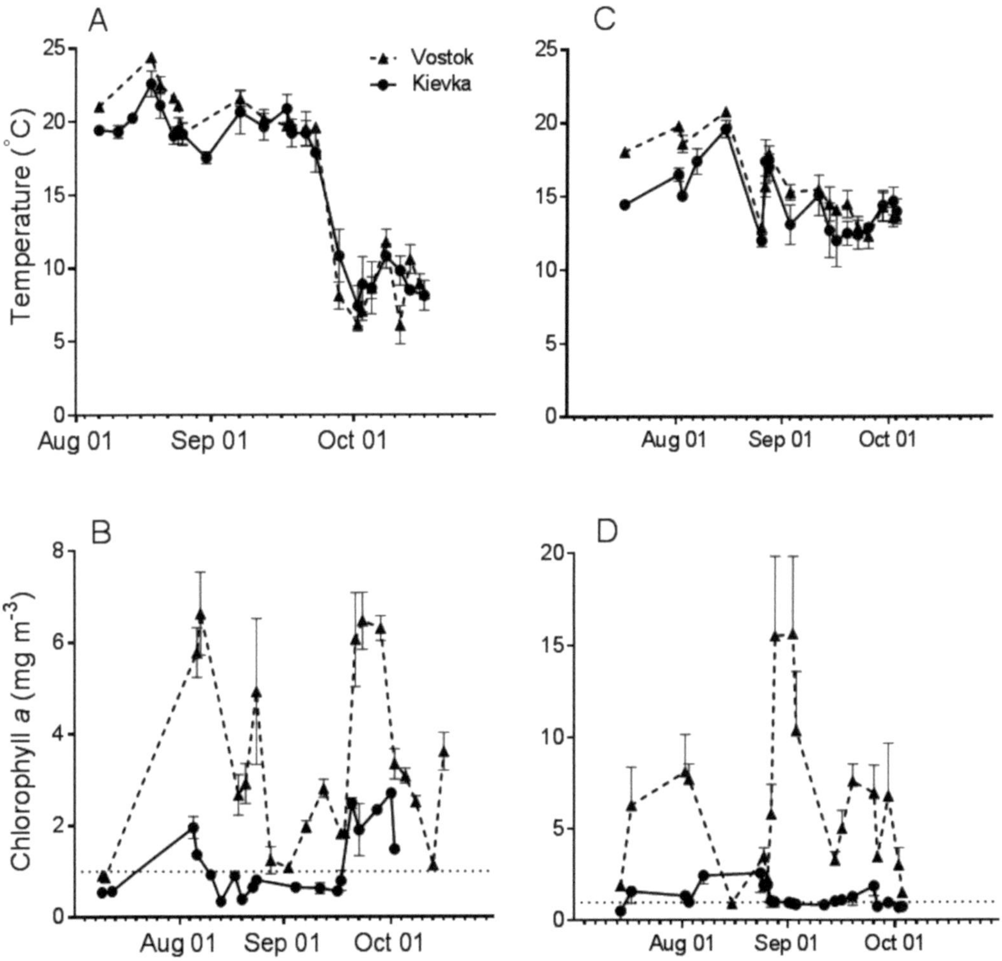

Figure 7. The satellite-based data (mean ± SD) showing temporal dynamics of the temperature (A, C) and Chl *a* concentration (B, D) in Vostok Bay and Kievka Bay (northwestern Sea of Japan) in summer–autumn seasons of 2008 and 2009. Horizontal dashed lines in (B, D) correspond to Chl *a* concentration of 1 mg m^{-3}. The data shown are extracted from **Figures 7–9** in Ref. [59].

In August–September 2009, temperature dynamics in Kievka and Vostok Bays (**Figure 7C**) as well as at other five sites (the Bays of Rudnaya, Vrangel, Shepalova, Sokolovskaya, and Kit, see **Figure 5** in Ref. [59]), was very similar. A sharp decrease in temperature due to upwelling was observed at all stations between August 28 and 31 followed by the second (lower) temperature oscillation. The proportions of females that failed to spawn, however, were different between different localities (**Figure 6**) and positively correlated with mean Chl *a* concentrations at these sites [59]. The highest Chl *a* concentration was observed in Vostok Bay (**Figure 7C**) where no unspawned females were found in 2009 (**Figure 6**).

All these data indicate that a link appeared to be between spawning success in *S. intermedius* populations and the phytoplankton level in the study area. In summer 2010, from July to October, we had the opportunity to measure the Chl *a* concentration, water temperature and salinity in Kievka Bay directly at the site of sea urchin sampling near the bottom. Analysis of gonadal state of the sea urchins was carried out at intervals of 8–14 days. Three subsequent spawning events were revealed between August 6 and September 8 based on significant decrease in the GI and a sharp increase in completely spawned sea urchins (**Figure 8A**). In 2010, sea urchin spawning was the most successful for over three years (2008–2010): in the sample collected on September 22, only one unspawned female of the 22 examined was found (**Figure 8A**).

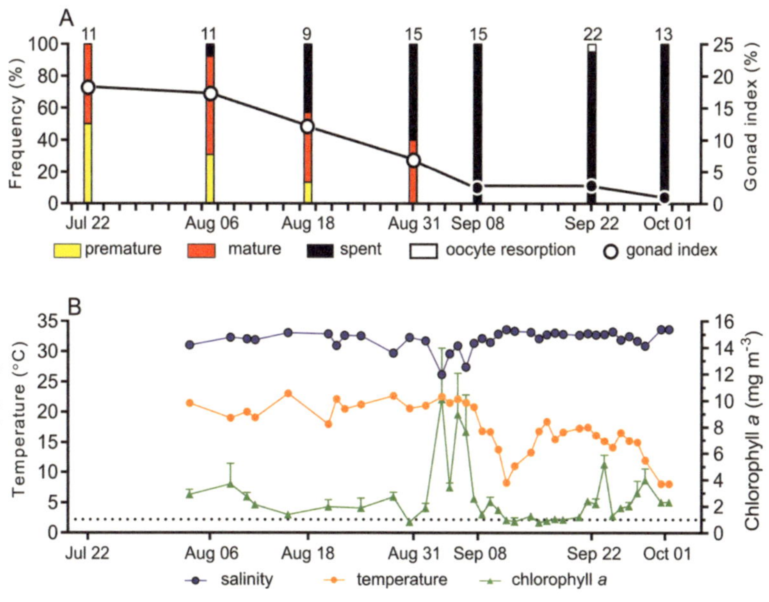

Figure 8. Temporal dynamics of gonadal development in *Strongylocentrotus intermedius* females (A) and environmental variables (B) in summer–autumn season of 2010 in Kievka Bay (northwestern Sea of Japan). (A) The percentages of the reproductive cycle stages (columns) and the GI values (circles and line; the figures above the columns show the number of sea urchins from which the frequency was calculated). (B) Temporal variation of the temperature, Chl *a* concentration and salinity (data of direct measurements in the bottom boundary layer). Horizontal dashed line corresponds to the Chl *a* concentration of 1 mg m^{-3}; solid vertical lines denote standard deviation. The data shown are compiled from **Figure 10B** and D in Ref. [59].

Between August 6 and August 18, 2010, when the first spawning event took place, the water salinity was relatively stable, whereas during August 18–31 and August 31–September 8, when the second and third spawning events occurred, a short-term decrease in salinity by 2–5‰ was recorded (**Figure 8B**). During the first and second spawning events, the temperature variations were from 17 to 23°C, whereas the third spawning event occurred under conditions of fairly stable temperature (**Figure 8B**). A sharp decline in temperature caused by upwelling occurred on September 11, when the sea urchin spawning was virtually completed. The Chl *a* concentration varied from 0.83 to 2.7 mg m^{-3} during the first and second spawning events and rose up to 10.1 mg m^{-3} during the third spawning event (**Figure 8B**).

Taken together, the results of our study indicate that a certain environmental stimulus or combination of stimuli is necessary for triggering the spawning in *S. intermedius* natural population. A sufficiently high concentration of phytoplankton, which serves as food for larvae, and water temperature suitable for larval development are the factors ensuring survival of sea urchin offspring; therefore, *S. intermedius* spawning may be triggered by these environmental stimuli. Our findings are not consistent with the studies presuming the changes in temperature may serve as proximal cues triggering the spawning of sea urchins (see for review, Ref. [46]). At the same time, the results of our analysis support a hypothesis that an increase in the phytoplankton concentration may be the most appropriate candidate for inducing sea urchin spawning.

Primary production in nearshore waters substantially depends on the influx of biogenic elements from the terrigenous runoff, upwelling and anthropogenic sources (see for review, Refs. [66–68]). The study area is characterized by spatially different level of anthropogenic pressure and significant year-to-year variations in the magnitude of river runoff and the upwelling schedule [56–59]. We hypothesized that the phenomenon of spawning failure in natural populations of *S. intermedius* seemed to be attributable to combination of environmental factors responsible for low primary productivity in water column during the sea urchin spawning season. The results of our subsequent studies showed that the phenomenon of spawning failure seems to be common in *S. intermedius* populations in Kievka Bay, so the sea urchin spawning season can last for more than 2 months [58], and spawning failure was observed during 4 reproductive seasons between 2008 and 2015 (**Figure 6**).

To test the hypothesis that phytoplankton induces or fails to induce spawning in *S. intermedius* populations, we conducted a detailed analysis of environmental parameters (concentrations of dissolved oxygen and Chl *a*, temperature, salinity, moon phases, and tide level) which were monitored directly in the animal habitats, along with examination of gonadal state of sea urchins sampled at a fine temporal scale (3–12 days, on average 4.2 days) [58].

4. Environmental variables influencing *S. intermedius* spawning

The studies were carried out in Kievka Bay at Sites 1–3 in 2011 and at Sites 1–4 in 2012 (**Figure 9**). An analysis of sea urchin gonads sampled at intervals of 3–12 days (on average, 4.2 days) was performed during the period from the beginning of August to the end of September (in the pre-spawning, spawning, and post-spawning periods of the reproductive cycle [59]).

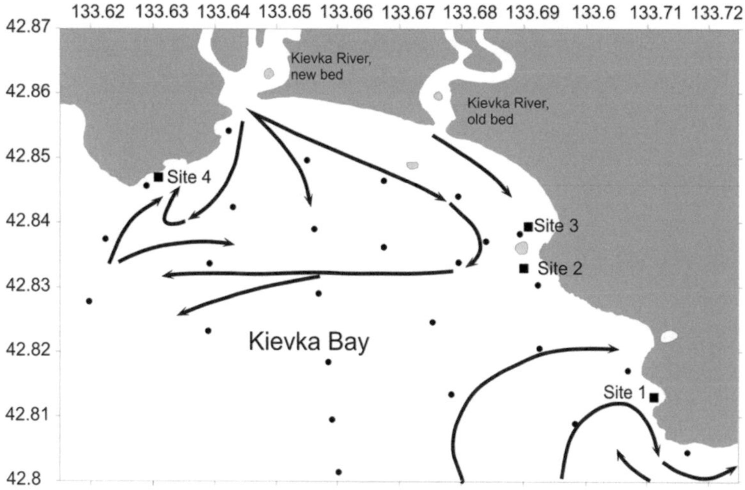

Figure 9. Map of Kievka Bay located in the northwestern Sea of Japan. The sites where sea urchins were collected are denoted by the squares, and 26 sites where the sounding of the water column was performed are denoted by small circles. Solid lines with arrowheads show surface currents in Kievka Bay in August (according to: Ref. [69]).

Fresh smears were used to determine the state of maturation of *S. intermedius* females. Six stages of ovarian maturation were determined: (1) immature, (2) premature, (3) partially mature, (4) mature, (5) spent, and (6) over-mature [58].

The measurement of depth, water temperature, salinity, photosynthetically active radiation, and concentrations of dissolved oxygen and Chl *a* was performed with a multi-parameter sonde directly at the sites of sea urchin sampling. At site 3, environmental variables were continuously (every 15 min) recorded by a YSI 6920V2 data logger which was installed 50 cm above the bottom. Additionally, these parameters were measured at 26 sites (**Figure 9**) in the bay by the CTD-sounding of the water column up to depth of 60 m. Details of the methods and statistical analysis are given in Ref. [58].

4.1. Gonadal maturity and spawning events

Examination of gonadal development (GI dynamics, ovarian maturity, and dynamics in the proportion of sea urchins ready for spawning) in *S. intermedius* at different sites in Kievka Bay throughout two reproductive seasons (August–October of 2011 and 2012) showed that sea urchins become mature by the August, so they have the potential to spawn at any time in August and most parts of September. However, mass spawning may be both synchronous and asynchronous between sea urchin populations within a bay; moreover, it was shown that not all populations spawned completely even within a comparatively small area. The differences in the dynamics of sea urchin spawning between the sites and the years are most evident from a comparison of the parameters of gonadal development in sea urchins from site 1 and those from other sites (**Figure 9**). In 2011, 30% of females and approximately 60% of males from site 1 failed to spawn which was evident from the analysis of the final sample taken on October 21 (**Figure 10A**). The percentages of females and males ready for spawning and mean GI value

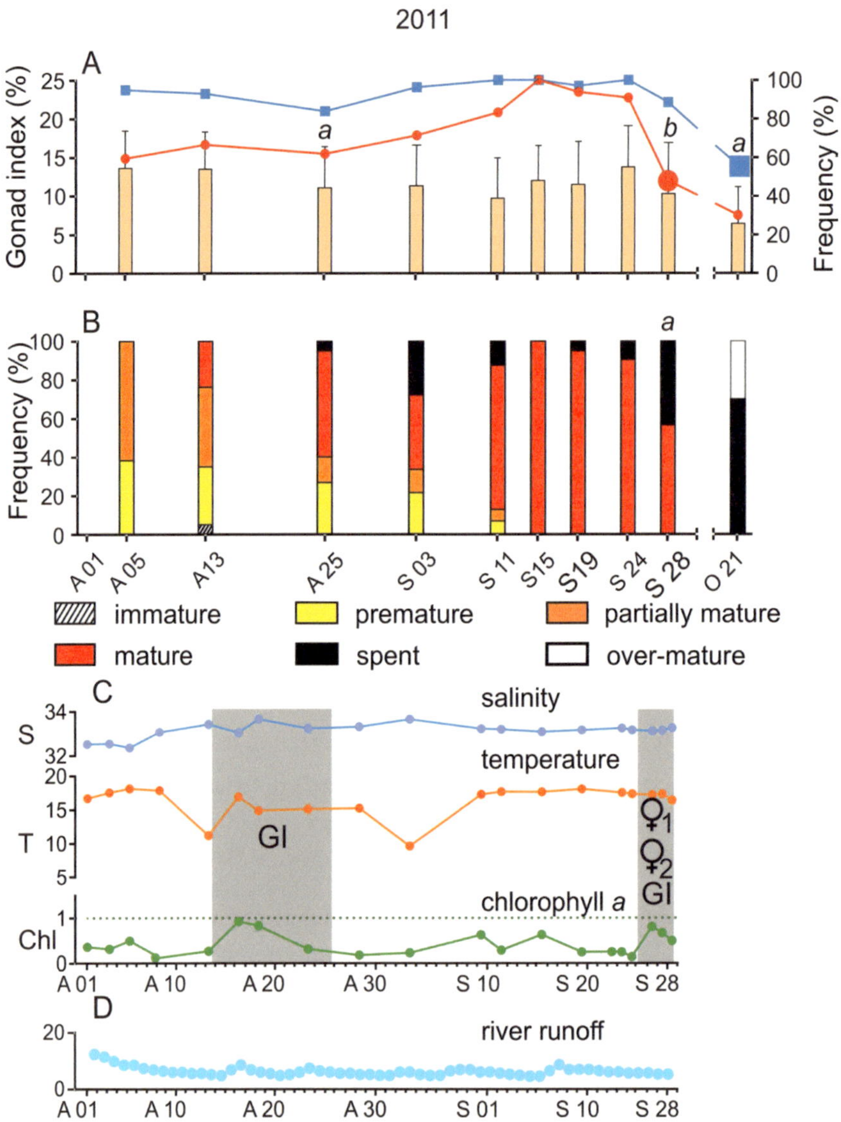

Figure 10. The timing of the spawning of the sea urchin *Strongylocentrotus intermedius* and the associated environmental conditions in August–September of 2011 at site 1 in Kievka Bay. (A) Temporal dynamics of the gonad index (columns: mean ± SD) and the percentages of the the males (lines with squares) and females (lines with circles) ready for spawning. Letters above the columns denote significant mean difference compared to previous month (*a*: $p < 0.05$, *b*: $p < 0.01$, one-way ANOVA, Kruskal-Wallis statistics, Dunn's multiple comparison test). Large squares and circles indicate significant differences compared to previous month ($p < 0.05$, binomial statistics). (B) Temporal dynamics of the percentages of the ovarian maturity stages in *S. intermedius* females. Letter "*a*" above the column denotes significant increase in the portion of spent females ($p < 0.05$, binomial statistics). (C) Spawning events and the dynamics of environmental variables: Chl—chlorophyll *a* (mg m^{-3}), S—salinity (psu), T—temperature (°C). Horizontal dashed line corresponds to the Chl *a* concentration of 1 mg m^{-3}. Shaded areas indicate the spawning windows determined by different parameters. GI: significant decrease in the gonad index; $♀_1$: significant decrease in the portion of females ready for spawning; $♀_2$: significant increase in the portion of spent females. (D) Daily runoff of the Lazovka River, the main tributary of the Kievka River that flows into Kievka Bay (m^3 s^{-1}). *x*-axis: month and date. The data shown are extracted from **Figures 2, 5,** and **7** in Ref. [58].

were rather a high due to a presence of unspawned individuals in the sample. Unspawned females had disintegrating egg in the ovaries, which corresponded to the over-mature stage (6) of the ovarian maturation (**Figure 10B**). Two sea urchin spawning events were revealed by a significant increase in the number of spent females and significant decrease in the GI and/or the percentage of females and males ready for spawning (**Figure 10C**, shaded areas). At the same time, at sites 2 and 3, the dynamics of the GIs, readiness for spawning, and level of ovarian maturity were significantly different from those at site 1 [58]. Three and two subsequent spawning events occurred during August–September, respectively, and all the females were spent on October 21. The GIs and percentages of females and males ready for spawning at sites 2 and 3 sharply decreased by September 28 and were close to zero on October 21.

In 2012, seasonal dynamics of the GIs, proportion of sea urchins ready for spawning and level of ovarian maturity at sites 1–4 were synchronous [58]. At site 1, more than a 50% decrease in the GI occurred between August 14 and August 18, and then, the GIs gradually decreased until October 17 up to values less than 5% (**Figure 11A**). At other three sites, the GIs exhibited very similar dynamics [58]. At all four sites, the seasonal dynamics of the proportions of females and males ready for spawning showed drops, mostly coincident with significant decreases in the GIs (see **Figure 11A**, for example). In September, the percentages of spent females at different sites constituted from 25 to 55% of the total female number (see **Figure 11B**, for example). From two to three sea urchin spawning events were revealed at different sites, and the analysis of the final sample taken on October 17 showed that the reproductive cycle of *S. intermedius* in all four populations studied ended in virtually complete spawning (see **Figure 11B** and **C**, for example).

In sum, in 2011 and 2012, 16 sea urchin spawning events (spawning windows) were revealed by a significant increase in the number of spawned females and significant decreases in the GI and/or the percentage of females ready for spawning.

4.2. Relationship between environmental variables and spawning events

The temporal patterns of *S. intermedius* spawning events throughout two reproductive seasons indicate that the exact timing of spawning during the spawning period depends on the local properties of the environment, that is, a certain external stimulus or a favorable combination of environmental conditions is necessary for spawning. Analysis of data logger measurements showed that the summer–autumn seasons of 2011 and 2012 differed in weather and hydrological conditions: (1) the quantity of precipitation and associated changes in the terrigenous input into Kievka Bay and bottom layer salinity were substantially higher in 2012 than in 2011; (2) seasonal wind-driven upwelling was more pronounced in 2011 than in 2012, this was responsible for the higher spatial and temporal variability in distribution of temperature in the water column in 2011 (see **Figures 10C** and **D** and **11C** and **D** for comparison). The oxygen content in the waters of Kievka Bay was close to saturation in both seasons.

To evaluate the relationships between temperature, Chl *a* concentration, salinity and the probability of sea urchin spawning, generalized linear models for binomial data (logistic regression) were used. The result of the registration of spawning events was used as a binary dependent variable, that is, a value of 1 was assigned to each inter-sample time interval when spawning

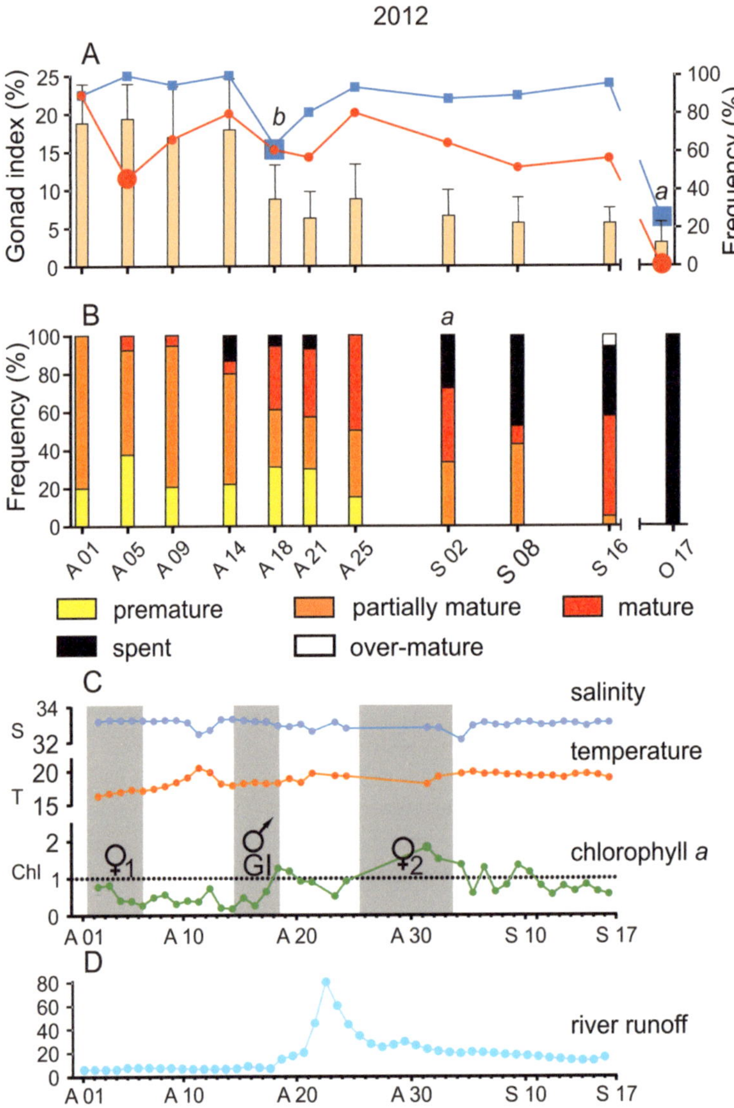

Figure 11. The timing of the spawning of the sea urchin *Strongylocentrotus intermedius* and the associated environmental conditions in August–September of 2012 at site 1 in Kievka Bay. (A) Temporal dynamics of the gonad index (columns: mean ± SD) and the percentages of the males (lines with squares) and females (lines with circles) ready for spawning. Letters above the columns denote significant mean difference compared to previous month (*a*: $p < 0.05$, *b*: $p < 0.01$, one-way ANOVA, Kruskal-Wallis statistics, Dunn's multiple comparison test). Large squares and circles indicate significant differences compared to previous month ($p < 0.05$, binomial statistics). (B) Temporal dynamics of the percentages of the ovarian maturity stages in *S. intermedius* females. Letter "*a*" above the column denotes significant increase in the portion of spent females ($p < 0.05$, binomial statistics). (C) Spawning events and the dynamics of environmental variables: Chl—chlorophyll *a* (mg m^{-3}), S—salinity (psu), T—temperature (°C). Horizontal dashed line corresponds to the Chl *a* concentration of 1 mg m^{-3}. Shaded areas indicate the spawning windows determined by different parameters. GI: significant decrease in the gonad index; $♀_1$: significant decrease in the portion of females ready for spawning; $♀_2$: significant increase in the portion of spent females. (D) Daily runoff of the Lazovka River, the main tributary of the Kievka River that flows into Kievka Bay (m^3 s^{-1}). *x*-axis: month and date. The data shown are extracted from **Figures 4, 6,** and **8** in Ref. [58].

occurred (spawning windows, n = 16), and a value of 0 was assigned to each inter-sample time interval when spawning did not occur (non-spawning windows, n = 55). Different combinations of environmental variables were used as continuous independent variables (predictors). The year of study or the site was used as a categorical independent variable. Both separate and combined influences of these predictors were analyzed using free and open-source R software [70].

No apparent relationship was revealed between temperature or salinity and the timing of spawning in *S. intermedius* populations. At the same time, our statistical analysis showed a significant positive relationship between Chl *a* concentration and spawning events in *S. intermedius* populations. These results provide strong support for the hypothesis that an increase in the concentration of phytoplankton may serve as the primary stimulus to trigger sea urchin spawning [13, 15–17, 71]. Based on the results of the laboratory experiments, it was proposed that phytoplankton stimulates the spawning of the most sensitive males, whose sperm promotes synchronous mass spawning. However, there are some contradictory results on the role of phytoplankton and sperm in inducing spawning. For example, phytoplankton alone did not induce spawning in experiments with the sea urchin *Lytechinus variegatus*, though the introduction of phytoplankton 75 min before sperm greatly accelerated the initiation of male spawning followed by female spawning [21]. In the field experiments, however, no consistent reaction of *L. variegatus* to sperm was observed [72]. The phenomenon of spawning failure repeatedly registered in *S. intermedius* populations from Kievka Bay during four reproductive seasons between 2008 and 2015 (**Figure 6**) indicates that *S. intermedius* males with mature gonads are not able to induce spawning in the absence of external stimuli. From a fine-scale temporal analysis of the dynamics of sea urchin gonadal development and environmental parameters, we can draw a conclusion that phytoplankton concentration is the main factor driving the initiation of *S. intermedius* spawning [58].

Our study showed that an increase in phytoplankton concentration in the bottom water layer of Kievka Bay, as judged from measured Chl *a*, occurred under different temperature conditions, that is, increasing, decreasing, or almost constant temperatures [58]. It is well known that warmer temperature promotes the acceleration of larval development of temperate sea urchin species, thereby decreasing the time when the larvae are susceptible to predation. The coincidence of increasing phytoplankton concentration and a stable warm or rising temperature, it seems, would benefit the reproductive success of *S. intermedius*. However, according to our data, such a combination of environmental factors occurs rather rarely in the study area. Our field observations showed that the spawning process in natural populations of *S. intermedius* lasted for approximately 1 h [58]. For this species, the time from fertilization to the formation of swimming blastulae over a temperature range of 15–20°C varied from 14 to 9.5 h [73]. We suggest a mechanism, involving a rapid response of *S. intermedius* parental individuals to increasing phytoplankton density, which promotes transportation of the sea urchin swimming larvae by advection together with warm water masses enriched with phytoplankton and thereby food supply for the offspring, even under conditions of the short-time oncoming of phytoplankton-rich warm water during a half-day tidal advection.

The next interesting finding of our study is a coincidence between the majority of *S. intermedius* spawning events and new and full moons [58]. This is in consistence with field observations of spawning events in a number of echinoids, both tropical (two species of the genus *Diadema* and two species of the genus *Echinothrix* [74]) and temperate (*S. droebachiensis* [16]), as well as with laboratory experiments on *L. variegatus* [21] which provide evidence of lunar periodicity in some species. However, the differently pronounced influence of the moon cycle on the spawning of *S. intermedius* in different years, as judged from different levels of both spawning success and synchronization between spawning events and lunar phases, forced us to suggest that this factor seems unlikely to be a proximal environmental cue for triggering *S. intermedius* spawning activity. Most likely, the lunar cycle may serve as an additional factor enhancing sea urchin sensitivity to other natural stimuli as proposed for *L. variegatus* [21].

5. Conclusion

In the northwestern Sea of Japan along 400 km of the coast of the Primorye region of Russia, three types of populations of the sea urchin *S. intermedius* were found that differ from each other in the proportions of individuals with early spawning (the end of May–June) and late spawning (September–early October). All *S. intermedius* populations studied live at the same latitude and experienced the same photoperiodic conditions, but the timing and implementation of spawning are different both between and within localities. The individuals with early spawning breed in spring and spawn in late spring–early summer under conditions of rising photoperiod (the duration of daytime) and temperature. However, the temperature in this period in the study area is much lower as compared to the warmest summer season when gonad maturation in the individuals with late spawning takes place. These individuals spawn under conditions of decreasing photoperiod and relatively stable warm temperature. Sharp temperature fluctuations in the study area caused by upwelling/downwelling do not drive *S. intermedius* spawning. Moreover, we found that, in some years, sea urchins from several localities failed to end their reproductive cycle with complete spawning despite similar temperature and photoperiod changes. Hence, neither photoperiod nor temperature may be considered as the primary external factors that determine *S. intermedius* temporal patterns of gonad maturation and the timing of spawning. We also found no apparent relationships between salinity, dissolved oxygen or tidal activity, and the spawning events.

Our analysis showed that since 1970–1980s, the shift in spawning season from autumn to early summer occurred in *S. intermedius* populations inhabiting anthropogenically polluted areas of Peter the Great Bay whereas sea urchins inhabiting relatively clean areas retained an autumn pattern of spawning. Based on these observations and on the results of the studies of population genetic structure of several *S. intermedius* populations with different spawning schedule which revealed no significant differences between the individuals with early and late spawning, we hypothesized that the shift of spawning seasonality from autumn to early summer would be a phenotypic switch related to the changes in environmental conditions caused by chronic anthropogenic pollution. Considering that (1) there is statistically significant relationship between *S. intermedius* spawning events and Chl *a* (phytoplankton) concentration; (2) spawning failure occurs in *S. intermedius* populations under conditions of low phytoplankton concentration throughout summer–autumn season; and (3) human activity is associated with the eutrophication of the seawater and phytoplankton increase, we propose that phytoplankton may be the main environmental factor driving reproductive cycle and spawning in *S. intermedius* populations. We hypothesize that permanent increased levels of phytoplankton and its metabolites in the eutrophic seawater would be environmental cues that are used by the sea urchins for assessing the appropriateness of environmental conditions for reproductive success in terms of benefits to offspring survival. We believe that *S. intermedius* reproductive cycle in the area not subjected to anthropogenic impact (Kievka Bay) may be considered as a "baseline" (reference) reproductive cycle for this species in the northwestern Sea of Japan. Analysis of the temporal patterns of *S. intermedius* gonadal maturation in Kievka Bay showed that a vast majority of individuals become mature by the end of July; however, approximately 5% of mature females and more than 50% of mature males were found in June [59]. Moreover, in June 2009, approximately 5% of spent individuals of both sexes were found. This indicates that *S. intermedius* possesses high reproductive plasticity which promotes the possibility of both early and late spawning depending on the local properties of the environment. Considering mentioned above, we can propose that there are at least two mechanisms responsible for the shift in *S. intermedius* timing of spawning from autumn to the early summer: (1) the seawater eutrophication promotes the probability of early spawning followed by the earlier beginning of a new wave of gametogenesis and gonad maturation and (2) sea urchin offspring from parents with early spawning pattern attains sexual maturity earlier than that from parents with late spawning pattern, thereby increasing the number of the individuals in sea urchin population which are able to spawn in late spring–early summer.

Acknowledgements

This work was partially supported by the grant of Russian Foundation for Basic Research No. 16-05-00083a and Russian Academy of Sciences Priority Program "The Far East" No.15-I-6-007o.

Author details

Peter M. Zhadan[1]*, Marina A. Vaschenko[2] and Tatyana N. Almyashova[1]

*Address all correspondence to: pzhadan@poi.dvo.ru

1 V. I. Il'ichev Pacific Oceanological Institute, Far Eastern Branch, Russian Academy of Sciences, Vladivostok, Russia

2 National Scientific Center of Marine Biology, A. V. Zhirmunsky Institute of Marine Biology, Far Eastern Branch, Russian Academy of Sciences, Vladivostok, Russia

References

[1] Giese AC. Comparative physiology: Annual reproductive cycles of marine invertebrates. Annual Review of Physiology. 1959;**21**:547-576. DOI: 10.1146/annurev.ph.21.030159.002555

[2] Giese AC, Pearse JS. Introduction: General Principals. In: Reproduction of marine invertebrates. Giese AC, Pearse JS, editors. Palo Alto, California: Blackwell Sci; 1974. pp. 1-49

[3] Sastry AN. Physiology and ecology of reproduction in marine invertebrates. In: Vernberg FJ, editor. Physiological Ecology of Estuarine Organisms. Columbia: University of South Carolina Press; 1978. pp. 279-299

[4] Lawrence AJ, Soame JM. The effects of climate change on the reproduction of coastal invertebrates. IBIS. 2004;**146**(Suppl. 1):29-39. DOI: 10.1111/j.1474-919X.2004.00325.x

[5] Levitan DR, Petersen C. Sperm limitation in the sea. Trends in Ecology & Evolution. 1995;**10**:228-231. DOI: 10.1016/S0169-5347(00)89071-0

[6] Yund PO. How severe is sperm limitation in natural population of marine free-spawners? Trends in Ecology & Evolution. 2000;**15**:10-13. DOI: 10.1016/S0169-5347(99)01744-9

[7] Levitan DR. Density-dependent sexual selection in external fertilizers: Variances in male and female fertilization success along the continuum from sperm limitation to sexual conflict in the sea urchin *Strongylocentrotus franciscanus*. American Naturalist. 2004;**164**:298-309. DOI: 10.1086/423150

[8] Guettaf M, San Martin G, Francour P. Interpopulation variability of the reproductive cycle of *Paracentrotus lividus* (Echinodermata : Echinoidea) in the south-western Mediterranean. Journal of the Marine Biological Association of the United Kingdom. 2000;**80**:899-907

[9] Brady SM, Scheibling RE. Changes in growth and reproduction of green sea urchins, *Strongylocentrotus droebachiensis* (Müller), during repopulation of the shallow subtidal zone after mass mortality. Journal of Experimental Marine Biology and Ecology. 2006;**335**:277-291. DOI: 10.1016/j.jembe.2006.03.016

[10] Marzinelli E, Bigatti G, Gimenez J, Penchaszadeh P. Reproduction of the sea urchin *Pseudechinus magellanicus* (Echinoidea: Temnopleuridae) from Golfo Nuevo, Argentina. Bull Mar Sci. 2006;**79**:127-136

[11] Agatsuma Y. *Strongylocentrotus intermedius*. In: Lawrence JM, editor. Sea Urchins: Biology and Ecology, 3rd edn. San Diego, CA: Academic Press; 2013. pp. 437–447

[12] Thorson G. Reproduction and larval development of Danish marine bottom invertebrates, with special reference to the planktonic larvae of the Sound (Øresund). Medd Komm Danm Fish Havunders Ser Plankton. 1946;**4**:1-529

[13] Himmelman JH. Phytoplankton as a stimulus for spawning in three marine invertebrates. Journal of Experimental Marine Biology and Ecology. 1975;**20**:199-214. DOI:10.1016/0022-0981(75)90024-6

[14] Smith JR, Strehlow DR. Algal-induced spawning in the marine mussel *Mytilus californicanus*. Invertebrate Reproduction and Development. 1983;**6**:129-133. DOI:10.1080/01651269.1983.10510032

[15] Starr M, Himmelman JH, Thermiault JC. Environmental control of green sea urchin, *Strongylocentrotus droebachiensis*, spawning in St Lawrence estuary. Canadian Journal of Fisheries and Aquatic Sciences. 1993;**50**:894-901. DOI: 10.1139/f93-103

[16] Gaudette J, Wahle RA, Himmelman JH. Spawning events in small and large populations of the green sea urchin *Strongylocentrotus droebachiensis* as recorded using fertilization assays. Limnology and Oceanography. 2006;**51**:1485-1496. DOI: 10.4319/lo.2006.51.3.1485

[17] González-Irusta JM, De Cerio FG, Canteras JC. Reproductive cycle of the sea urchin *Paracentrotus lividus* in the Cantabrian Sea (northern Spain): Environmental effects. Journal of the Marine Biological Association of the United Kingdom. 2010;**90**:699-709. DOI: 10.1017/S002531540999110X

[18] Starr M, Himmelman JH, Therriault JC. Direct coupling of marine invertebrate spawning with phytoplankton blooms. Science. 1990;**247**:1071-1074. DOI: 10.1126/science.247.4946.1071

[19] Breese WP, Robinson A. Razor clams, *Siliqua patula* (Dixon): Gonadal development, induced spawning and larval rearing. Aquaculture. 1981;**22**:27-33. DOI: 10.1016/0044-8486(81)90130-7

[20] Starr M, Himmelman JH, Therriault JC. Isolation and properties of a substance from the diatom *Phaeodactylum tricornutum* which induces spawning in the sea urchin *Strongylocentrotus droebachiensis*. Marine Ecology: Progress Series. 1992;**79**:275-287

[21] Reuter KE, Levitan DR. Influence of sperm and phytoplankton on spawning in the echinoid *Lytechinus variegatus*. Biological Bulletin. 2010;**219**:198-206. DOI: 10.1086/BBLv219n3p198

[22] Guillou M, Lumingas LJL. The reproductive cycle of the 'blunt' sea urchin. Aquaculture International. 1998;**6**:147-160. DOI: 10.1023/A:1009290307840

[23] Himmelman JH, Dumont CP, Gaymer CF, Vallières C, Drolet D. Spawning synchrony and aggregative behaviour of cold-water echinoderms during multi-species mass spawnings. Marine Ecology: Progress Series. 2008;**361**:161-168. DOI: 10.3354/meps07415

[24] Wilson BR. Survival and reproduction of the mussel *Xenostrobus securis* (Lamarck) (Mollusca; Bivalvia; Mytilidae) in a western Australia Estuary. Pt. II: Reproduction, growth and longevity. Journal of Natural History. 1969;**3**:93-120. DOI: 10.1080/00222936900770111

[25] Bazhin AG. The sea urchin genus *Strongylocentrotus* in the seas of Russia: Taxonomy and ranges. In: Mooi R, Telford M, editors. The 9th International Echinoderm Conference; 1998 August 1996; San Francisco. Rotterdam: Balkema; 1998. pp. 563-566

[26] Kafanov AI, Pavlyuchkov VA. Ecology of the commercial sea urchins genus *Strongylocentrotus* off continental Japan Sea. Proceedings of the Pacific Research Institute of Fisheries and Oceanography (TINRO). 2001;**128**:349-373 (in Russian with English summary)

[27] Vashchenko MA, Zhadan PM, Almyashova TN, Kovalyova AL, Slinko EN. Assessment of the contamination level of bottom sediments of Amursky Bay (Sea of Japan) and their potential toxicity. Russian Journal of Marine Biology. 2010;**36**:359-366. DOI: 10.1134/S1063074010050056

[28] Lutaenko KA, Vaschenko MA. Amursky Bay—An ecosystem under stress. In: Lutaenko KA, Vaschenko MA, editors. Ecological Studies and the State of the Ecosystem of Amursky Bay and the Estuarine Zone of the Razdolnaya River (Sea of Japan). Vladivostok: Dal'nauka; 2008. pp. 7-29

[29] Vashchenko MA. Pollution in Peter the Great Bay, Sea of Japan, and its biological consequences. Russian Journal of Marine Biology. 2000;**26**:155-166. DOI: 10.1007/BF02759533

[30] Vaschenko MA, Pitruk DL. Comparative assessment of the condition of the marine environment and biota in various parts of Peter the Great Bay including the Tumen River mouth area. In: Kasyanov VL, Vaschenko MA, Pitruk DL, editors. The State of Environment and Biota of the Southwestern Part of Peter the Great Bay and the Tumen River Mouth. Vladivostok: Dal'nauka; 2001. pp. 3-18

[31] Moshchenko AV, Belan TA, Oleinik EV. Effect of contamination on soft bottoms in Peter the Great Bay, Sea of Japan: The choice of variables. In: Lutaenko KA, Vaschenko MA, eds. Ecological Studies and the State of the Ecosystem of Amursky Bay and the Estuarine Zone of the Razdolnaya River (Sea of Japan). Vladivostok: Dal'nauka; 2009. pp. 223-266

[32] Long ER, MacDonald DD. Recommended uses of empirically derived, sediment quality guide lines for marine and estuarine ecosystems. Human and Ecological Risk Assessment 1998;**4**:1019-1039. DOI: 10.1080/10807039891284956

[33] Fuji A. Studies on the biology of the sea urchin. III. Reproductive cycle of two sea urchins, *Strongylocentrotus nudus* and *S. intermedius*, in southern Hokkaido. Bulletin of the Faculty of Fisheries Hokkaido University. 1960;**11**:43-48

[34] Matsui T, Agatsuma Y, Ogasawara M, Taniguchi K. Coincidence in reproduction of the sea urchin *Strongylocentrotus intermedius* in Hirota Bay, on the Pacific Ocean off Northern Honshu, and in the Sea of Japan off Hokkaido, Japan. Journal of Shellfish Research. 2008;**27**:1283-1289. DOI: 10.2983/0730-8000-27.5.1283

[35] Zhadan PM, Vashchenko MA, Almyashova TN. The study of the nature of desynchronization of the reproductive cycle of the sea urchin *Strongylocentrotus intermedius* in the northwestern part of the Sea of Japan. Herald of the Far East Branch of the Russian Academy of Sciences. 2011;**4**:40-48 (in Russian with English summary)

[36] Gnezdilova SM. Morphological and cytochemical characteristics of oogenesis and reproductive cycles of the sea urchins *Strongylocentrotus nudus* and *S. intermedius* [thesis]. Vladivostok: Institute of Marine Biology; 1971 (in Russian)

[37] Yakovlev SN. The seasons of reproduction of the sea urchins *Strongylocentrotus nudus* and *S. intermedius* in Vostok Bay, the Sea of Japan. In: Kasyanov VL, editor. Biological Studies of Vostok Bay. Vladivostok: Far East Science Center, USSR Academy of Sciences; 1976. pp. 136-142 (in Russian)

[38] Khotimchenko YS, Deridovich II, Motavkin PA. Biology of Reproduction and Regulation of Gametogenesis in Echinoderms. Moscow: Nauka; 1993. 168 p (in Russian)

[39] Kasyanov VL, Medvedeva LF, Yakovlev YM. Reproduction of the Echinoderms and Bivalves. Moscow: Nauka; 1980. 280 p (in Russian)

[40] Voropaev VM, Strakhov AA. The spring season of maturation in *Strongylocentrotus intermedius*. Russian Journal of Zoology. 1977;**568**:1260-1262 (in Russian with English summary)

[41] Vaschenko MA, Zhadan PM, Latypova EV. Long-term changes in the state of gonads in sea urchins *Strongylocentrotus intermedius* from Amur Bay, the Sea of Japan. Russian Journal of Ecology. 2001;**32**:358-364. DOI: 10.1023/A:1011982412642

[42] Vaschenko MA, Almyashova TN, Zhadan PM. Long-term and seasonal dynamics of the gonad state in the sea urchin *Strongylocentrotus intermedius* under anthropogenic pollution (Amursky Bay, Sea of Japan). Herald of the Far East Branch of the Russian Academy of Sciences. 2005;**1**:32-42

[43] Ouréns R, Fernández L, Freire J. Geographic, population, and seasonal patterns in the reproductive parameters of the sea urchin *Paracentrotus lividus*. Marine Biology. 2011;**158**:793-804. DOI: 10.1007/s00227-010-1607-1

[44] Lozano J, Galera J, Lopez S, Turon X, Palacin C, Morera G. Biological cycles and recruitment of *Paracentrotus lividus* (Echinodermata: Echinoidea) in two contrasting habitats. Marine Ecology: Progress Series. 1995;**122**:179-191. DOI: 10.3354/meps122179

[45] Soualili D, Guillou M. Variation in the reproductive cycle of the sea urchin *Paracentrotus lividus* in three differently polluted locations near Algiers (Algeria). Marine Biodiversity Records. 2009;**2**:1-6. DOI: 10.1017/S175526720900092X

[46] Mercier A, Hamel J-F. Endogenous and exogenous control of gametogenesis and spawning in echinoderms. Advances in Marine Biology. 2009;**55**:1-302. DOI: 10.1016/S0065-2881(09)55008-0

[47] Agatsuma Y, Kawamata K, Motoya S. Reproductive cycle of cultured seeds of the sea urchin, *Strongylocentrotus intermedius* produced from geographically separated population. Suisanzoshoku. 1994;**42**:63-70 (in Japanese with English summary)

[48] Agatsuma Y, Momma H. Release of cultured seeds of the sea urchin, *Strongylocentrotus intermedius* (A. Agassiz), in the Pacific coastal waters of southern Hokkaido. I. Growth and reproductive cycle. Scientific Reports of Hokkaido Fisheries Experimental Station. 1988;**31**:15-25 (in Japanese with English summary)

[49] Zaslavskaya NI, Vashchenko MA, Zhadan PM. The genetic structure of populations of the sea urchin *Strongylocentrotus intermedius* from the northwestern Sea of Japan in connection with a shift in spawning time. Russian Journal of Marine Biology. 2012;**38**:325-338. DOI: 10.1134/S1063074012040104

[50] Bondar EI, Tatarenkov A, Avise JC. Population genetic structure of the commercial sea urchin *Strongylocentrotus intermedius* from Peter the Great Bay and north-eastern coasts of the Primorsky region. In: Modern Achievements in Population, Evolutionary, and Ecological Genetics: International Symposium; September 2-6, 2013; Vladivostok-Vostok Marine Biological Station. Program & Abstracts.Vladivostok: IMB FEB RAS; 2013. p. 17

[51] Ebert TA. Adaptive aspects of phenotypic plasticity in echinoderms. Oceanologica Acta. 1996;**19**:347-355

[52] Boudouresque CF, Verlaque M. Ecology of *Paracentrotus lividus*. In: Lawrence JM, editor. Edible Sea Urchins: Biology and Ecology, 2nd edn. Amsterdam: Elsevier; 2007. pp. 243-285

[53] Calderon I, Giribet G, Turon X. Two markers and one history: Phylogeography of the edible common sea urchin *Paracentrotus lividus* in the Lusitanian region. Marine Biology. 2008;**154**:137-151. DOI: 10.1007/s00227-008-0908-0

[54] Aleksanin AI, Kim V, Orlova TY, Stonik IV, Shevchenko OG. Phytoplankton of the Peter the Great Bay and its remote sensing problem. Oceanology (Mosc). 2012;**52**:219-230. DOI: 10.1134/S0001437012020014

[55] Stonik IV, Orlova TY. Summer-autumn phytoplankton in Amurskii Bay, Sea of Japan. Russian Journal of Marine Biology. 1998;**24**:207-213

[56] Zhabin IA, Gramm-Osipova OL, Yurasov GI. Wind-induced upwelling off the north-western coast of the Japan Sea. Russian Meteorology and Hydrology. 1993;**10**:82-86 (in Russian with English summary)

[57] Zvalinskii VI, Lobanov VB, Zakharkov SP, Tishchenko PY. Chlorophyll, delayed fluorescence, and primary production in the northwestern part of the Sea of Japan. Oceanology (Mosc). 2006;**46**:23-32. DOI: 10.1134/S0001437006010048

[58] Zhadan PM, Vaschenko MA, Lobanov VB, Sergeev AF, Kotova SA. Fine-scale temporal study of the influence of hydrobiological conditions on the spawning of the sea urchin *Strongylocentrotus intermedius*. Marine Ecology: Progress Series. 2016;**550**:147-161. DOI: 10.3354/meps11678

[59] Zhadan PM, Vaschenko MA, Almyashova TN. Spawning failure in the sea urchin *Strongylocentrotus intermedius* in the northwestern Sea of Japan: Potential environmental causes. Journal of Experimental Marine Biology and Ecology. 2015;**465**:11-23. DOI: 10.1016/j.jembe.2014.12.017

[60] Holland ND, Giese AC. An autoradiographic investigation of the gonads of the purple sea urchin (*Strongylocentrotus purpuratus*). Biological Bulletin. 1965;**128**:241-258

[61] Reunov AA, Kalachev AV, Yurchenko OV, Au DWT. Selective resorption in nutritive phagocytes of the sea urchin *Anthocidaris crassispina*. Zygote. 2004;**12**:71-73. DOI: 10.1007/s00441-004-0901-y

[62] Reunov AA, Yurchenko AV, Kalachev AV, Au DWT. An ultrastructural study of phagocytosis and shrinkage in nutritive phagocytes of the sea urchin *Anthodicaris crassispina*. Cell and Tissue Research. 2004;**318**:419-428. DOI: 10.1017/S0967199403002429

[63] Walker CW, Harrington LM, Lesser MP, Fagerberg WR. Nutritive phagocyte incubation chambers provide a structural and nutritive microenvironment for germ cells of *Strongylocentrotus droebachiensis*, the green sea urchin. Biological Bulletin. 2005;**209**:31-48. DOI: 10.2307/3593140

[64] Walker CW, Unuma T, Lesser MP. Gametogenesis and reproduction of sea urchins. In: Lawrence JM, editor. Edible Sea Urchins: Biology and Ecology, 2nd edn. Amsterdam: Elsevier; 2007. pp. 11-33

[65] Masuda RI, Dan JC. Studies on the annual reproductive cycle of the sea urchin and the acid phosphatase activity of relict ova. Biological Bulletin. 1977;**153**:577-590. DOI: 10.2307/1540607

[66] Hecky RE, Kilham P. Nutrient limitation of phytoplankton in freshwater and marine environments: A review of recent evidence on the effects of enrichment. Limnology and Oceanography. 1988;**33**:796-822. DOI: 10.4319/lo.1988.33.4part2.0796

[67] Howarth RW. Coastal nitrogen pollution: A review of sources and trends globally and regionally. Harmful Algae. 2008;**8**:14-20. DOI: 10.1016/j.hal.2008.08.015

[68] Cloern JE, Foster SQ, Kleckner AE. Review: Phytoplankton primary production in the world's estuarine-coastal ecosystems. Biogeosciences Discussions. 2013;**10**:17725-17783. DOI: 10.5194/bg-11-2477-2014

[69] Zuenko Y. Fisheries Oceanography of the Japan Sea. Vladivostok: Publishing House of the Pacific Research Institute of Fisheries and Oceanography (TINRO-Center); 2008. 227 p (in Russian)

[70] R Development Core Team. R: A Language and Environment for Statistical Computing [Internet]. Vienna: R Foundation for Statistical Computing; 2010. Available from: www.R-project.org [Accessed: February 21, 2017]

[71] Levitan DR. Density-dependent selection on gamete traits in three congeneric sea urchins. Ecology. 2002;**83**:464-479. DOI: 10.1890/0012-9658(2002)083[0464:DDSOGT]2.0.CO;2

[72] McCarthy DA, Young CM. Effects of water-borne gametes on the aggregation behavior of *Lytechinus variegatus*. Marine Ecology: Progress Series. 2004;**283**:191-198. DOI: 10.3354/meps283191

[73] Buznikov GA, Podmarev VK. Sea urchins *Strongylocentrotus droebachiensis*, *S. nudus*, and *S. intermedius*. In: Dettlaff TA, Vassetzky SG, editors. Animal Species for Developmental Studies Invertebrates. New York: Consultants Bureau; 1990. pp. 253-285

[74] Coppard SE, Campbell AC. Lunar periodicities of diadematid echinoids breeding in Fiji. Coral Reefs. 2005;**24**:324-332. DOI: 10.1007/s00338-005-0491-5

Section 3

Aquaculture as a Tool Against the Depletion of Sea Urchin Stocks

Chapter 4

Morphological and Biochemical Profiles of the Gonadal Cycle in the Sea Urchin *Paracentrotus lividus*: Wild Type vs. Bred

Bernardetta Anna Tenuzzo, Elisabetta Carata,
Stefania Mariano and Luciana Dini

Additional information is available at the end of the chapter

http://dx.doi.org/10.5772/intechopen.68895

Abstract

Paracentrotus lividus gonads represent a valued gourmet delicacy, particularly appreciated in Europe and in Japan. Their commercial value is generally associated to their size, freshness, colour and texture. Diet, gametogenesis and environmental conditions have a marked influence, promoting the indispensable mechanisms of synthesis, selective storage and mobilization of the bioactive compounds, as lipids, proteins and carbohydrates of gonads in order to obtain nutrients. The objective of this work is to compare the morphological and biochemical profiles of reproductive life cycle of the gonads of adult *P. lividus* in its marine natural environment and adult captured sea urchins breeding into a fish aquaculture system. The reproductive cycle of male and female wild and breeding *P. lividus* was characterized during 1 year by analysing variations of the gonadal content of lipids, proteins and carbohydrates of animals captured at four different locations of the south-western coast of Salento, Italy, with the animals grown in a fish farm and fed with four different types of diet. The gonadal and repletion indexes were determined before the specimen dissection for evaluation of sex, development stages and physiological aspects. Gonads were processed for histological and biochemical analysis. The gonadal content of lipids, proteins and carbohydrates was performed by the gas chromatography-mass spectrometry (GC-MS) and by spectrometry, respectively.

Keywords: sea urchin, gonadal cycle, lipids, proteins, carbohydrates

1. Introduction

The potential marketing of the edible sea urchin *Paracentrotus lividus* (Lamark 1816) has been considerably augmented during the last two decades due to its culinary value for various Mediterranean and mid-European populations. This fact poses particular interest for this invertebrate species [1] that is the most studied group of benthic macrofauna in the Mediterranean sea [4]. This edible Mediterranean sea urchin species has an extensive geographical distribution in Apulia sea, in the south of Italy [2]. However the local consumption of edible sea urchin is more substantial, coinciding with the peak of the tourist season. Local populations of sea urchins experience unavoidable pressure during that season, and in some cases, local authorities have imposed a temporary or permanent ban to protect them. Several studies have reported that sampling activities (fishing, harvesting) have direct consequences on the benthic macrofauna, potentially decreasing density and individual size and leading to sex ratio imbalance [3]. Furthermore, because edible sea urchins are frequently found in shallow waters, they are also subject to recreational fishing [4, 5].

The valuable food product of sea urchins is the gonads of female animals, referred as *roe*. The largest fisheries are in Japan, Chile and the USA, the latter two countries primarily exporting the product to markets in the Far East. The Chilean urchin fishery has increased dramatically in recent years to around 54,000 tons, largely due to the discovery of new fishing grounds. Smaller fisheries exist also in Europe, mainly supplying domestic markets. However the fishery statistics clearly demonstrate that most of the world's urchin fisheries are fully or over-exploited, and it is generally accepted that further urchin fishing grounds are unlikely to be discovered. Whereas the biological basis for sea urchin culture has been long established, research continues to refine the hatchery production of the juveniles [6, 7]. To promote sea urchin production, different strategies, relatively to the optimal diets for sea urchin, in order to improve the quality of gonads (colour, taste and texture) are reported [8–11]. For example, in Japan the cages are allocated, immediately (juvenile period) in sea floor, but in other countries, the cages are first incubated in area with a low quantity of nutrient and later are arranged in sea floor [8–11]. Hatchery-reared juveniles have been grown in suspended culture with *Atlantic salmon*, *Salmo salar* [12–14]; in closed recirculation systems [15]; in land-based integrated systems [16] and in rock pools in southern Ireland (J. Chamberlain, Dunmanas Seafoods Ltd., pers. Comm.). Another interesting and cost-effective option is to obtain a uniform quality of gonads combining the advantages derived from the systems that produce a uniform size class, with the systems that use the coculture [17]. In north-west Scotland, *P. lividus* is being evaluated as a potential new species for aquaculture. *P. lividus* is often described as herbivorous, although there are documented instances of its feeding on artificial diets containing fish meal [18], sponges, hydrozoa, copepods, dead fishes and mussels [19]. Diet quality can significantly influence the sea urchins' somatic and gonadal growth [20–24].

Aquaculture has been supporting human demands for fish products for centuries and is an important industry worldwide [25]. Aquaculture fisheries are a booming industry but discharging heavy nutrient loads into coastal water [24] and on an intensive scale causing severe environmental problems. Modern intensive monoculture requires high inputs of water, feeds, fertilizers and chemicals and inevitably produces considerable wastes. Therefore, many aquaculture operations

http://dx.doi.org/10.5772/intechopen.68895

put enormous pressure on coastal habitats [26]. Waste products from fish farms consist mainly of nitrogen, phosphorus and carbon dioxide. A possible solution to this problem is to integrate seaweeds or filtrating species into fish farming. Numerous studies have been performed which combine seaweed culture with land-based fish tanks or open sea fish cages [24, 27–29]. Seaweeds removed up to 90% of the nutrients discharged from an intensive fish farm. Algal farming along the coasts, therefore, may function as an effective biofilter to alleviate the eutrophication problem worldwide [28–33]. Moreover, the use of filtrating species has also been considered in an integrated cocultured system and showed reasonably high efficiency in the removal of waste inorganic nutrients [24–29]. In this respect, the present research work is aimed to set up the best diet for the optimal breeding conditions of *P. lividus* in a coculture offshore fish farm. To this purpose the morphological and biochemical characteristics of sea urchin grown in their natural marine environment and in fish farm were compared. The wild samples were captured in four different sites along the Ionian sea of Apulia and near the offshore fish farm located in Torre Suda. The sampling sites were chosen: two, Capilungo and Posto Rosso, at south and two, Arcobaleno and Torre Pizzo, at north of fish farm (**Figure 1**). These animals were analysed during 1 year for the monitoring of the growth by comparing the population living in natural conditions, in relation to the sea flowing variability of available food with the population of sea urchins in breeding conditions and fed with different determined diets (**Figure 1**).

1.1. Gonad quality: the importance of lipids, protein and carbohydrates

Gonadal growth and maturation of sea urchins are characterized by an accumulation of nutrients in the nutritive phagocyte cells that are then used for gametogenesis. Pre-gametogenesis in both sexes begins with the gonad size increase by accumulation of nutrients into the phagocytic cells, filling the gonadal lumina. During gametogenesis, nutritive phagocytes gradually decrease in size, supplying nutrients to developing germ cells; lumina progressively are filled with eggs and sperms [34, 35]. Diet, gametogenesis and water temperature may have a marked influence on gonad quality, promoting the indispensable mechanisms of synthesis, selective storage and

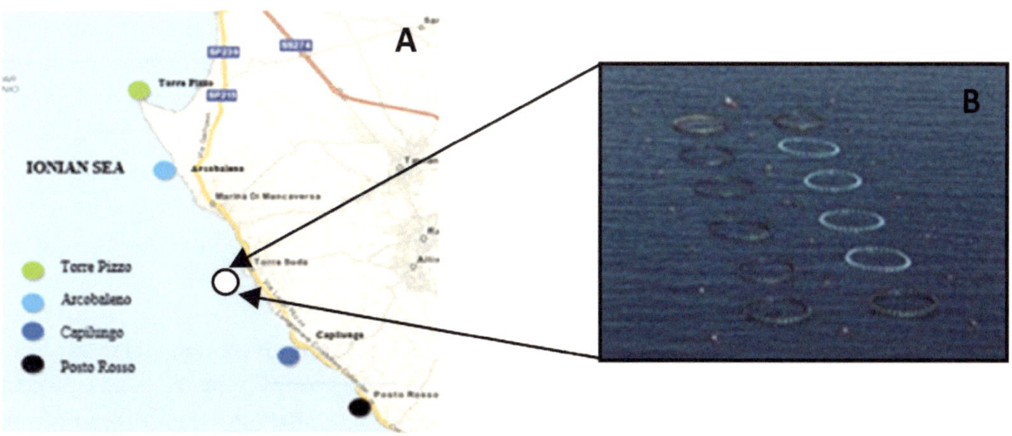

Figure 1. (A) Map of Ionian Salento coast with the sampling sites of wild sea urchins. Location of fish farm is indicated by the circle; (B) overview of the cage distribution in the offshore aquaculture system at Torre Suda, Gallipoli, Italy.

contents [36–39]. Indeed, quality and quantity of food affect the reproductive maturation and growth of sea urchins [20–39] and biochemical composition of gonads [40–43]. Lipids, proteins and carbohydrates are all necessary for the sea urchin growth. Their presence in the gonads allows the correct gametogenesis.

Lipids, important energy reserves storing more energy per unit volume than proteins or carbohydrates, are structural components of cell and subcellular membranes and are vital for somatic growth. Lipids, like fatty acids (FA) and polyunsaturated fatty acids (PUFAs), are present in the gonads and are essential for a multitude of physiological functions. In addition PUFAs have also significant effects on human health, like eicosapentaenoic acid (EPA, C20:5 (n−3)) and docosahexaenoic acid (DHA, C22:6 (n−3)), that can prevent arrhythmia, cardiovascular diseases and cancer [44, 45]. A large group of FAs characterized by a furan ring (FFA) [46, 47], whose activity in nature is not completely clear, show a promising nutritional value, with scavenger activity against hydroxyl and peroxyl radicals [46]. The composition of these valuable FFAs, however, varies greatly among different urchin species and is influenced by their natural diet as well as physiological processes, that is, reproductive stage [18, 35]. FAs are also important for sea urchin reproduction. During gametogenesis, they can be used as a source of energy [48], and additionally, sea urchin spermatozoa obtain energy for swimming through oxidation of fatty acids derived either from phosphatidylcholine or from triglycerides [49]. The latter are important for larval development and survival [50].

Proteins are one of the most necessary and costly nutrients of most aquatic animal diets. Adequate provision of dietary protein decreases feed intake and increases growth and roe production in all species of sea urchin. On the other hand, high protein levels or the presence of specific amino acids could have an adverse effect on the quality of sea urchin roe. Therefore, an adequate protein diet should be formulated to get maximal growth and roe production, avoiding protein excess.

Soluble carbohydrates are easily digested by sea urchins, and numerous carbohydrates have been identified in the sea urchin gut [35], indicating that sea urchins can most likely utilize carbohydrates from a wide array of sources. Few studies have examined the relationship between dietary protein and dietary energy requirements in sea urchins. Understanding this relationship may be an important step in the formulation of a feed suitable for sustainable sea urchin aquaculture.

2. Materials and methods

2.1. Sample management

The study was carried out from July 2015 to June 2016 in a fish farm located 900 m far from the south coast of the Gallipoli (Lecce, Italy) area in the Ionic Sea (**Figure 1**). Four different cages are fed with artificial diet which was composed of a mixture in equal proportion of organic maize kernel (Zea mays), previously crushed with a blender into grains of a few millimetres, and chopped fresh organic spinach leaves (*Spinacia oleracea*); soya-based diet; *Ulva lactuca* fresh-based diet and pellet diet Classic K (Hendrix S.p.A). The diets were administered ad libitum; all cages are bred in coculture with *Sparus aurata* and *Dicentrarchus labrax*. The wild-type samples were collected monthly, in the same period, in four different localities: at south (Capilungo and Posto Rosso) and north (Arcobaleno and Torre Pizzo) of the fish farm. *P. lividus* are situated at 5–6 m depth. It is a steep zone and the bottom is mainly rocky. The diet of wild type was chiefly based on seaweed *U. lactuca*.

2.2. Biometric measures and histological studies

For each sampling site, 50 animals were collected randomly and brought back to the laboratory alive. Measurements of the "height" along the oral-aboral axis, in mm "diameter", perpendicular to the oral-aboral axis and "wet weight" in grams were taken for each animal. The gonadosomatic index (GSI) was calculated as GSI = 100 × wet weight of gonads/wet weight of whole animal. For the estimation of a feeding index, the gut contents of specimens per sample were dry-weighed and used to calculate a repletion index (RI) as proposed by Kempf [53], RI = 100 × wet weight of gut/wet weight of whole animal [51]. Specimens were then dissected to collect gonads that were weighted and the stage was determined. All the individuals with a gonad dry weight less than 4 g were considered to be immature or undifferentiated. Gonads were fixed in Bouin's liquid for histological studies. Samples were dehydrated in alcohol and embedded in agar-paraffin wax, and 2 μm sections were stained with hematoxylin and eosin [2]. Maturity of the gonads was estimated according to stages established by Lozano et al. [51] and Sanchez-Espana et al. [52], which included four phases of development in the female and two in the males.

2.3. Biochemical composition

2.3.1. Lipid extraction and quantification

Fatty acids were methylated with a modified version of the method proposed by Antongiovanni et al. [54]. The total lipids were extracted from fresh sample (gonads and gut) five times with hexane solution of internal standards, containing 1 mg/ml each of methyl valerate, methyl nonanoate, methyl tridecanoate and methyl nonadecanoate. Hence, 40 μl of methanolic KOH 2 N was added. The mixture was vortexed for 1 min at room temperature, and then the hexane phase was analysed by GC. The product of each extraction was filtered and then collected in one round glass flask.

Methylations were carried out in triplicate for each extract. Mixtures of fatty acid methyl esters (FAMEs) were analysed using a modified version of the method proposed by Santercole et al. [55], using one temperature programme with a 175°C plateau. Each FAME was identified using pure FAME reference materials, a custom reference mixture and Menhaden fish oil analytical standard (Sigma Aldrich, St. Louis, MO, USA). All chemicals and solvents were of analytical grade and were purchased from Sigma Aldrich (St. Louis, MO, USA). The sample was dried under vacuum by Rotavapor (Buchi) and the residue was dissolved with toluene. The transesterification reaction was performed with BF3 in methanol for 20 min at 90°C. Then by adding water and shaking vigorously, we obtained two phases, only the organic phase, containing lipids derivatized, was analysed by Gas Chromatografy with Flame Ionization Detector (GC-FID) for quantitative analysis.

2.4. Quantification of protein

To measure the protein content, 0.2 g of the gonad was homogenized with 3 ml of radioimmunoprecipitation assay buffer (RIPA buffer) and centrifuged at 3500 g for 10 min. Protein in the solution was measured by the Bradford method [56], with bovine serum albumin as a standard.

2.5. Quantification of polysaccharides

The polysaccharide content was extracted according to Unuma 2002. Polysaccharide in the solution was measured according to the anthrone-sulfuric acid method [57] with glucose as a standard.

3. Results

3.1. Gonadal growth and quality

The gonad maturity of female and male sea urchins was observed for 1 year in wild and breeding animals. Sea urchins have a reproductive cycle characterized by (i) a growing stage, the nutrients are accumulated in the nutritive phagocytes; (ii) a maturation stage, the nutrients are transferred to germ cells for gametogenesis and lastly (iii) a spawning stage, mature gametes are released from the gonad. The gonadal annual cycle for female distinguishes four stages: H1, H2, H3 and H4, while the gonadal cycle for male shows three stages M0, M1 and the spawning identified as M1s. In **Figure 2**, the histological analysis of gonadal cycle for female sea urchins is reported. In the first stage (H1), said *turned off* stage, ascini are devoid of gametes, the ascinal walls are thin and the lumen is filled with nutritive phagocytes. The primary oocytes are few along the ovary wall. The second stage (H2), or *oocytes maturation* phase, coincides with the onset of vitellogenesis and the consequent size increase of the primary oocytes. The oocytes remain attached to the ovarian capsule walls and are still surrounded by nutritive phagocytes. Primary oocytes of larger dimension start to migrate towards the middle of the ovary capsule and displace nutritive phagocytes. In the third stage (H3), which is the *spawning and post-spawning* stage, the ovaries are filled with eggs, and the nutritive phagocytes form a thin layer along the capsule ovary wall. The last stage (H4) consists in the gonadal reconstitution with a thick nutritive layer. In the capsule of the ovaries, the number of mature eggs is reduced, and empty spaces are observed due to post-spawning. In addition, the wall of the ovary is almost devoid of cells and is very thin. The histological analysis of female gonads from wild sea urchins grown in the different sites and those breeding did not show histological differences, while a temporal variability of gonadal cycle distribution during the year was observed (**Table 1**).

The stage M0, or *turned off* stage, of male gonadal cycle is characterized by lack of sperms and by nutritive phagocytes at the periphery. During the stage M1, the presence of sperm is observed, the mature testes are packed with spermatozoa and the nutritive phagocytes are absent or limited to the periphery. Males' specimens showed a similar pattern in all locations.

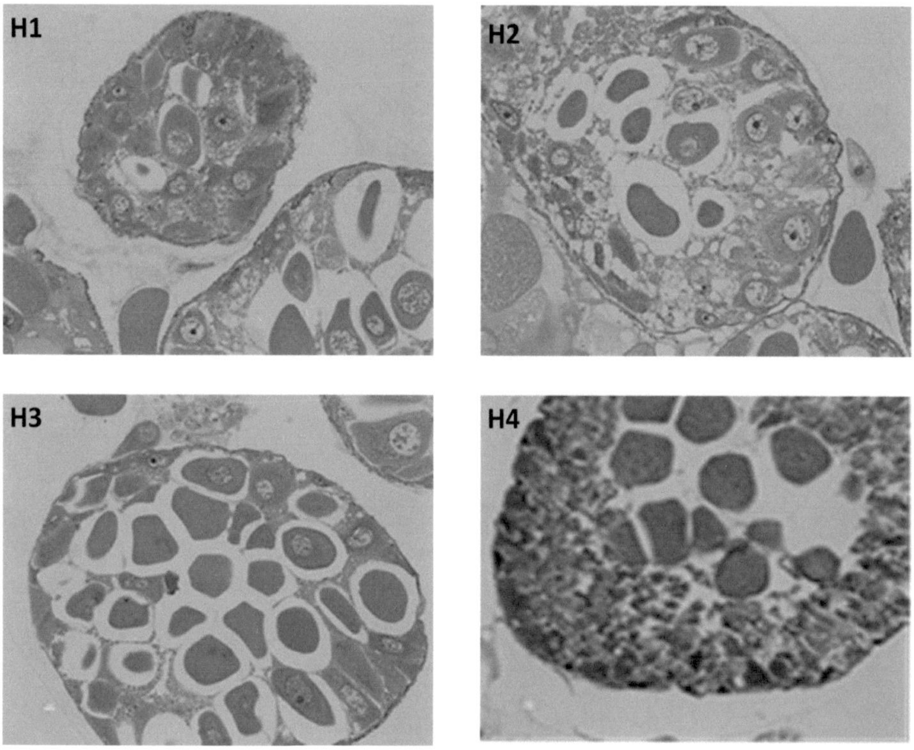

Figure 2. Histological representation of wild-type female gonad phases, that is, H1, H2, H3 and H4.

	H1	H2	H3	H4
Different sites				
Capilungo	Oct–Dec	Jan–Mar	Apr–Jul	Aug–Sept
Posto Rosso	Sep–Dec	Jan–Apr	May–Jul	Aug–Sept
Torre Pizzo	Sep–Nov	Dec–Feb	Mar–Jun	Jul–Aug
Arcobaleno	Sep–Nov	Dec–Feb	Mar–May	Jun–Aug
Breeding condition				
U. lactuca diet	Oct–Nov	Dec–Feb	Mar–Jun	Jul–Sep
Pellet-based diet	Sep–Dec	Jan–Mar	Apr–Jun	Jul–Aug
Soya-based diet	Sep–Dec	Jan–Apr	May–Jun	Jul–Aug
Artificial diet	Oct–Dec	Jan–Mar	Apr–May	Jul–Sept

Table 1. Annual distribution of gonadal cycle phases in female sea urchins.

The histological analysis of male gonads from wild sea urchins grown in the different sites and those in bred conditions shown a temporal variability of gonadal cycle distribution during the year (**Table 2**) (**Figure 3**).

	M0	M1	M1s
Different sites			
Capilungo	Oct–Mar	Apr–Jun	Jul–Sep
Posto Rosso	Sep–Mar	Apr–Jun	Jul–Aug
Torre Pizzo	Nov–Mar	Apr–Aug	Sep–Nov
Arcobaleno	Nov–Apr	May–Aug	Sep–Oct
Breeding condition			
Ulva lactuca diet	Oct–Feb	Mar–May	Jun–Sep
Pellet-based diet	Sep–Mar	Apr–Jul	Aug
Soya-based diet	Sep–Feb	Mar–May	Jun–Aug
Artificial diet	Nov–Apr	May–Jun	Jul–Oct

Table 2. Annual distribution of gonadal cycle phases in male sea urchins.

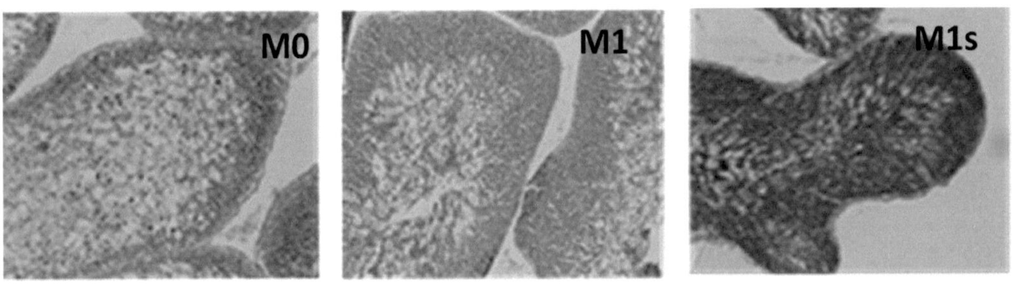

Figure 3. Histological representation of wild-type male gonads phases, that is, M0, M1 and M1s.

The gonadosomatic index (GSI) vs. size of wild animals captured at the four sites and of the animals fed with four different diets in the fish farm are shown in **Figure 4**. The lowest GSI values were observed in 15–35 mm size class collected in Capilungo, Posto Rosso and Arcobaleno, with Arcobaleno being the lowest one. In all four cages, the GSI of 15–35 mm size class of sea urchins are comparable to Torre Pizzo. A significant rise of GSI was observed in the 36–50 mm size class for all sites. High GSI values were measured for the 51–60 mm size class of Capilungo, while in Arcobaleno, the GSI value is comparable between 36–50 class and 51–60 class sizes. In Posto Rosso and Torre Pizzo, the GSI decreased significantly in the largest size class. In sea urchin cages, the GSI values of 36–50 mm size class were higher than for the wild conditions except for the animals fed with a soya-based diet.

The greatest repletion index (RI) values were found in the 51–60 mm size class in all four sites (**Figure 5**). The same trend was observed for the different sites: RI values increased with the increase of the size of sea urchins. Significant differences were found among the northern sites and southern sites: Capilungo and Posto Rosso measured the highest values for the 36–50 mm size class. The temporal trend of RI highlights the differences between the stations at north of Torre Suda (i.e. Capilungo and Posto Rosso) and those at south of Torre Suda (i.e.

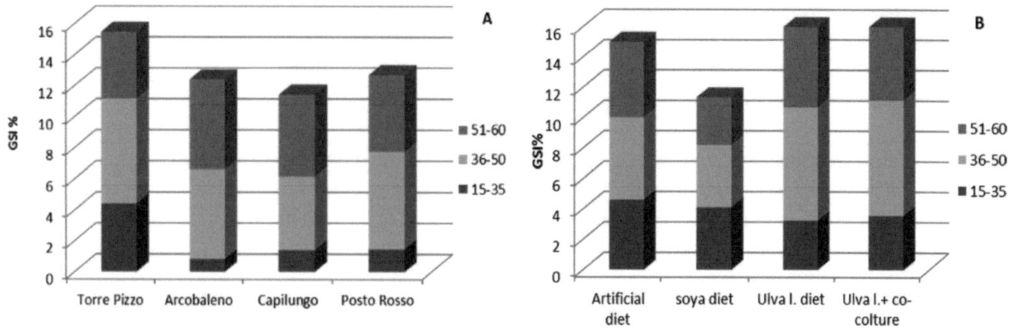

Figure 4. The gonadosomatic index (GSI) in wild and breeding animals.

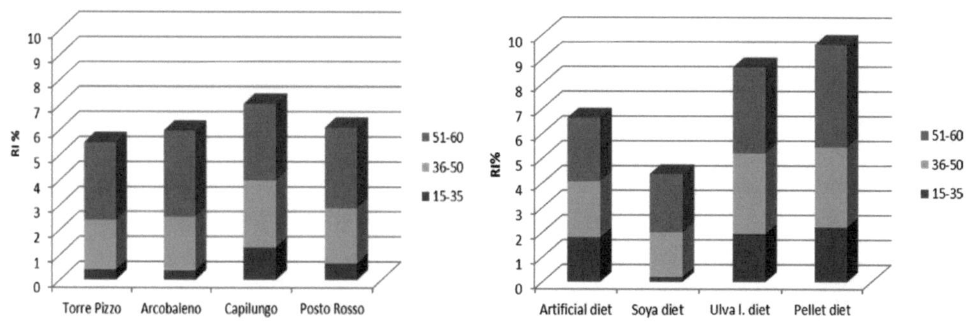

Figure 5. The repletion index (RI) in natural and in breeding conditions.

Arcobaleno and Torre Pizzo) in terms of values. In absolute, the highest RI values were measured in August in Capilungo and Posto Rosso, while the highest RI values for Arcobaleno and Torre Pizzo were measured in March. In the breeding sea urchins, the same trends of the environmental condition of Capilungo and Posto Rosso were observed; also in the breeding condition, the highest value was observed in August.

3.2. Biochemical profiles

Quantitative changes in the content of proteins, lipids and polysaccharides are analysed in wild samples (Capilungo, Posto Rosso, Torre Pizzo and Arcobaleno) and in breeding samples (four different diets). The content of proteins, lipids and polysaccharides decreased with the gametogenesis ongoing. The quantitative change of each class of molecules in the gonads during gametogenesis was standardized to the sizes of gonads. **Figures 4–7** show changes in the biochemical composition of ovaries and testes in function of phase of gonadal cycle (**Tables 1** and **2**) standardized to g100 g^{-1} body mass. Protein level in female gonads remained relatively constant between phases H1 and H2 (September and February) and decreased to a minimum value during phase H4 (July and August) in wild samples. Also in female gonads of breeding sea urchins, protein level remained relatively constant during the phases H1–H2 (October and March) and decreased to a minimum value during phase H4 (July and August).

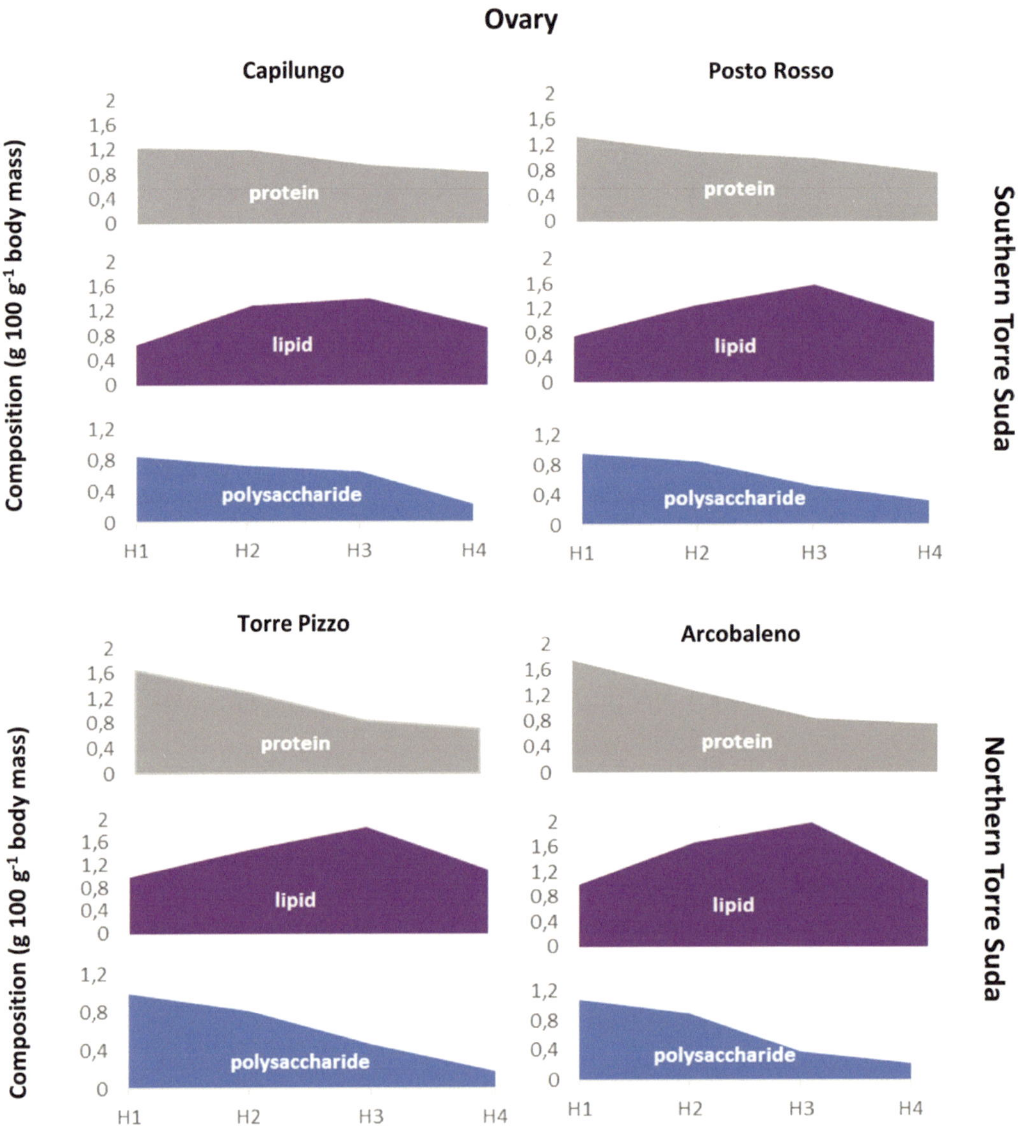

Figure 6. Biochemical composition of ovary of *P. lividus* wild type.

In wild male gonads, the content of protein is dependent on the site of sampling; in fact the maximum level of proteins in the wild samples grown at south of fish farm is registered during phase M1 (June); instead, the high value of proteins are registered in August at north of fish farm. In the breeding samples, the maximum content of proteins is recorded at the end of phase M1 (over spring/summer). The carbohydrate content remained constant in wild-type and breeding female gonads over phases H3 and H4, respectively (July and until June). The wild male gonads sampled at south of fish farm show a swift decrease in June, while the decrease was observed in August for the sample sites at north of fish farm. For the breeding samples, the speedy decrease of carbohydrate content is recorded during phase M1 (spring season). The

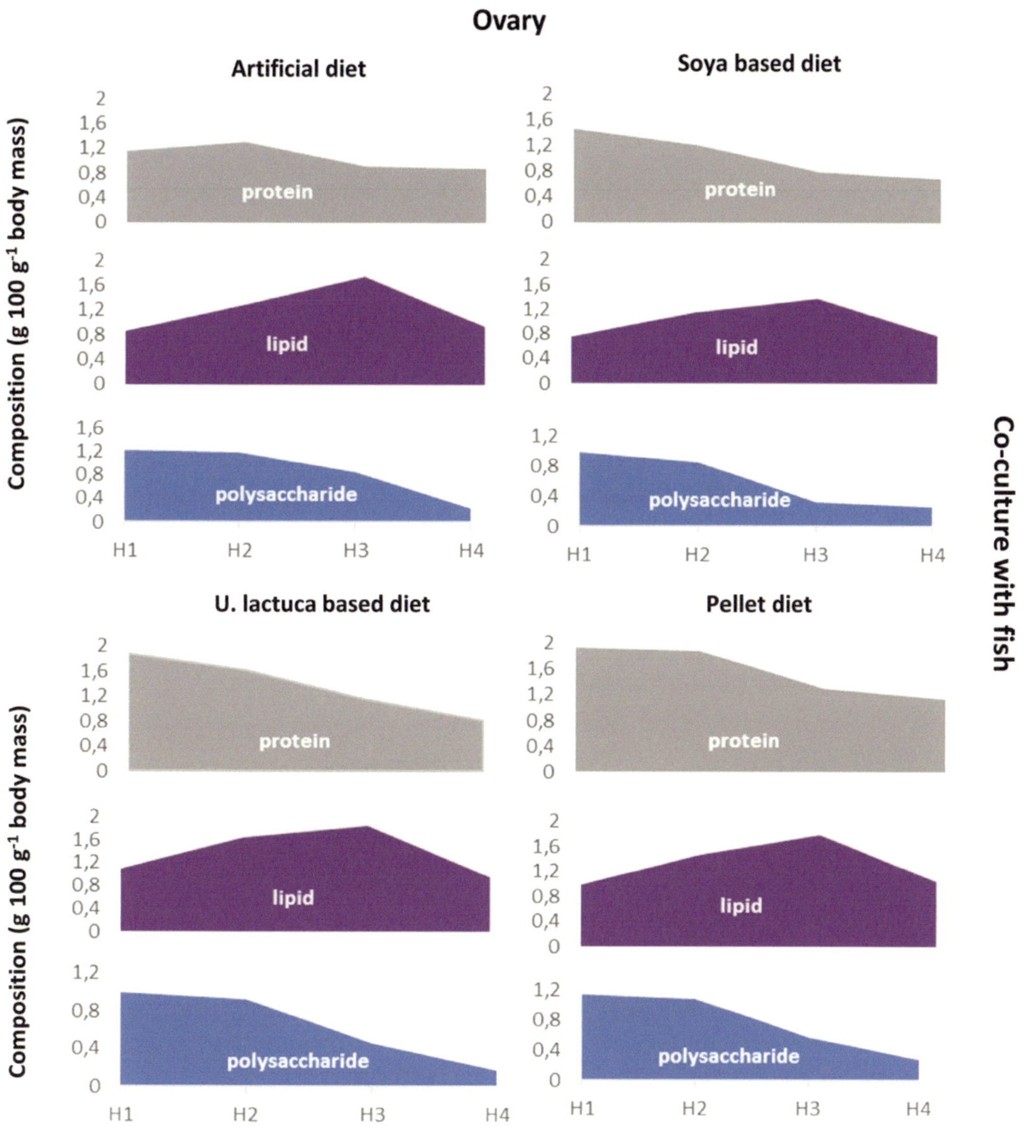

Figure 7. Biochemical composition of ovary of *P. lividus* bred.

lipid content underwent monthly variations in sea urchin gonads. In contrast to proteins and carbohydrates, the lipids are relatively growing until the start of phase H4 (July–August) in both wild and in bred female animals; afterwards the lipid content decreases simultaneously with the release of gametes. In the male gonads, the content of lipids was maximum during the phase M1 (spring season) for both wild and bred animals and decreased after the spawning.

In the light of these observations, it turns out that the process of gonadal maturation of breeding animals is retarded of about 1–2 months compared to the wild type. Moreover, there are also differences among the four groups of animals fed with different diets; the best content in lipids, proteins and polysaccharides is found in cages fed with pellets and *U. lactuca*, followed by cages fed with artificial diet. Animals raised on a soy-based diet have a lower lipid profile than other breeding animals and wild. Artificial diet consumption was comparable to those recorded for the most preferred algae, whereas maize consumption was significantly higher than those referred to pellet. This feeding behaviour could be due to the high content of carbohydrates and proteins in maize with respect to algae and spinach which exhibit a very similar biochemical composition, characterized by a moderate content of protein and lipid intake (**Figures 8** and **9**).

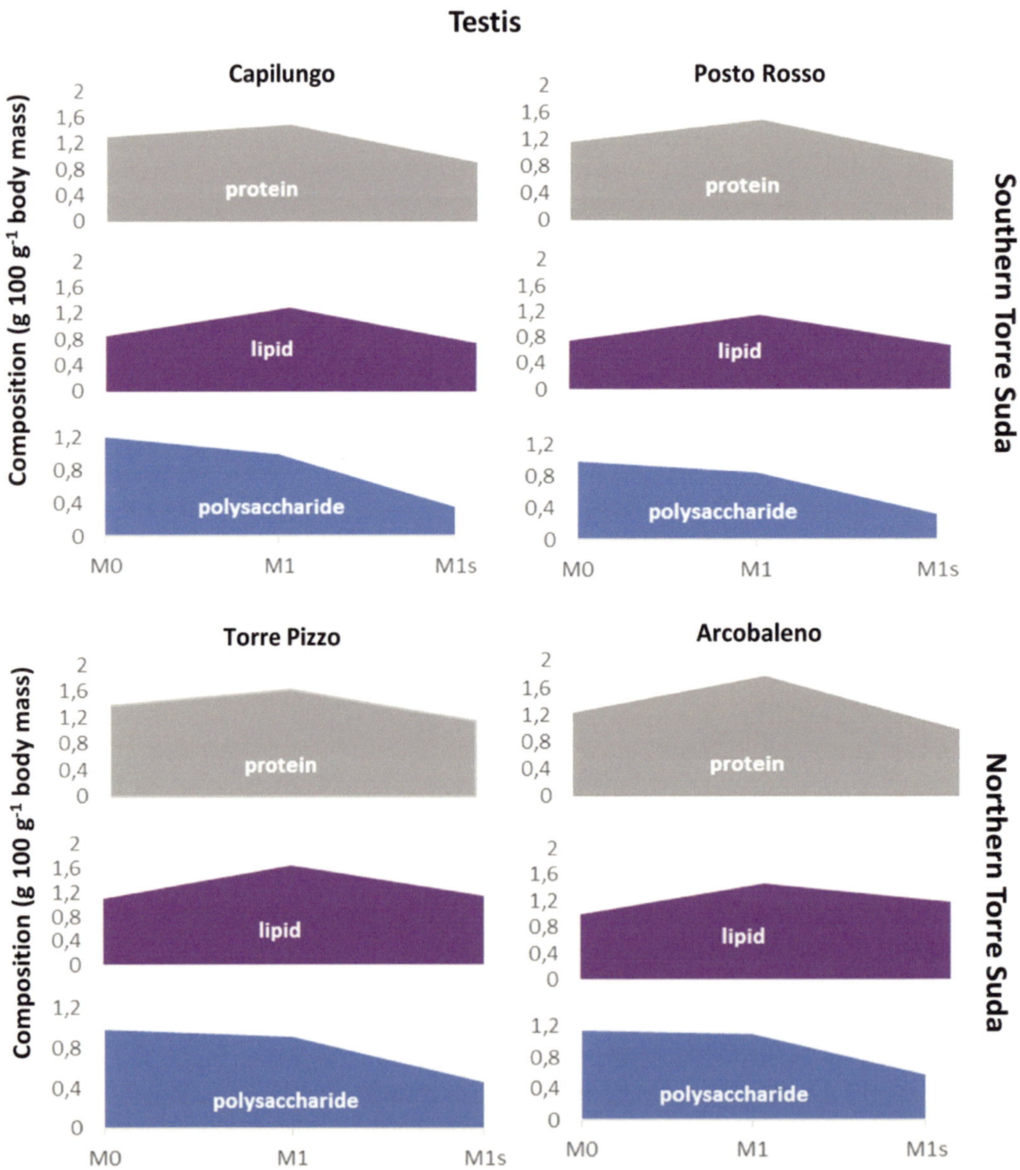

Figure 8. Biochemical composition of the testis of *P. lividus* wild type.

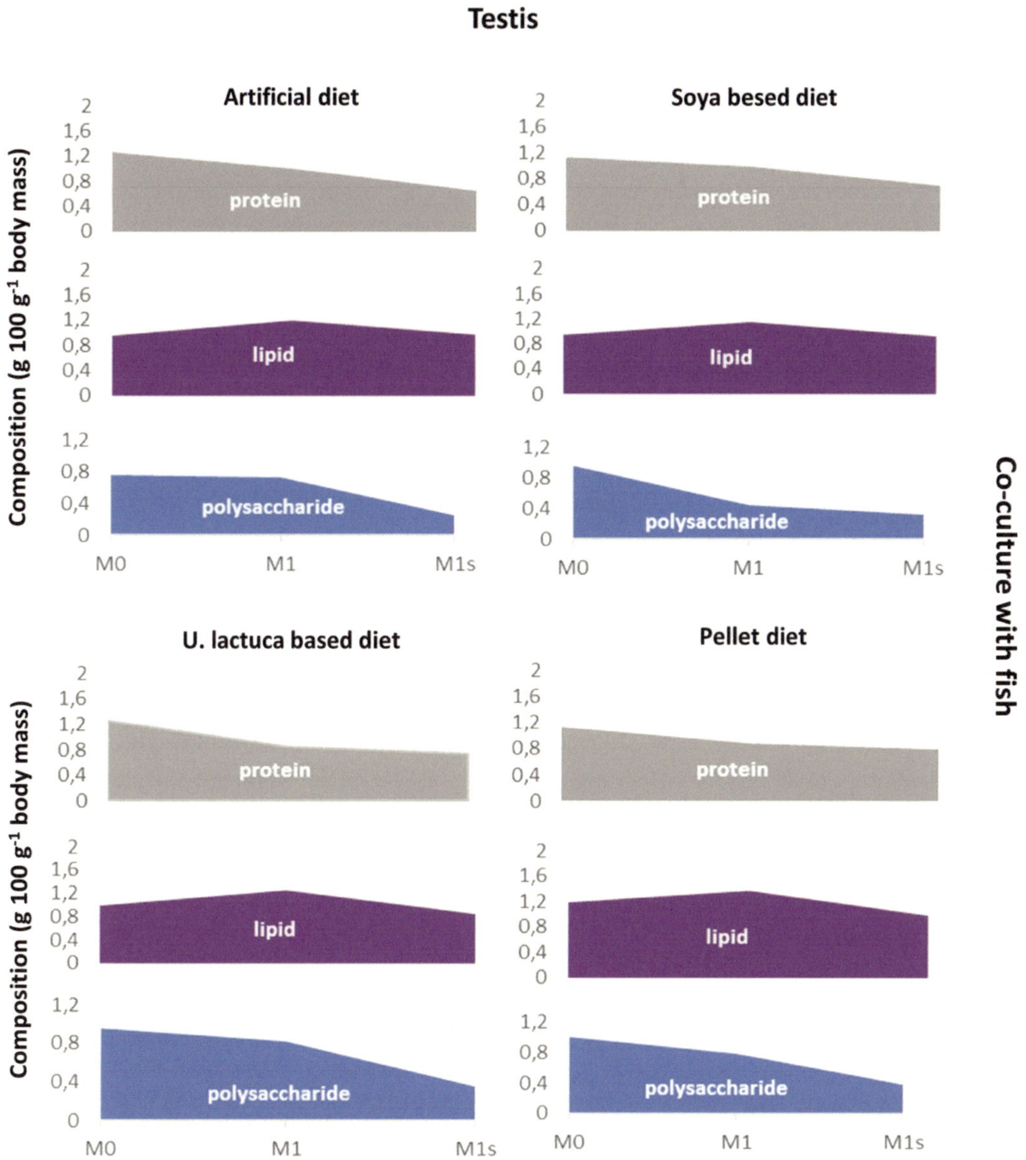

Figure 9. Biochemical composition of the testis of *P. lividus* bred.

4. Discussion

In the present work, the morphological and biochemical gonad characteristics of sea urchin grown in their natural marine environment and in cocultures with fish in an offshore fish farm were analysed for a period of 1 year. These data, which allows to define the best diet for sea urchins in view of the breeding set-up conditions, represent an important starting point for a scale production of sea urchins. Sea urchins are luxury food, and thus there is a growing economical interest for their cultivation, but *P. lividus* is also central for its importance in determining the structure of rocky reefs [58]. Taking into consideration the fact that *P. lividus* is one of the most intensely harvested benthic invertebrate species for commercial and recreational purpose, the definition of optimal conditions to set up sea urchin culture is thus pivotal from broad perspectives. Our data demonstrated the good feasibility of a low-cost and easy-to-standardize diet, such as based on fresh *U. lactuca* and artificial in rearing vulnerable species, such as *P. lividus* in aquaculture system. The specimens of sea urchin demonstrated a good propensity to ingest both maize and spinaches, content in artificial diet, as it can be inferred from the RI reported in **Figure 5**; in addition, the GSI values related to this diet are the highest registered in this study. This diet also shows a low mortality rate, about 20%, compared to the other diets that bring a maximum value of mortality as 30% (data not shown). Other researchers have described the benefit of artificial diets over the use of maize: Repolho assessed the effect of captive brood-stock diet on fertilization and endotrophic larvae development of *P. lividus* obtained for maize diet [59, 60]. Histological analysis revealed a good quality of gonadal tissue in all cages, but compared to the wild animals, the gonadal cycle seems to be slowed down by 1–2 months, despite the biochemical profile of the gonads that does not reveal alterations in the amount of the different components. The biochemical composition of the sea urchin gonads has been studied in different species; the immature ovary and testis contained a big quantity of polysaccharides, lipids and proteins. In our study the polysaccharides decrease during the gametogenesis in both sexes and in wild and bred animals; according to literature these macromolecules are involved in the energetic metabolism [35, 48]. During the gametogenesis the lipids are stable for long time and decrease in the late phases of gonadal cycle for both sexes. Fatty acids, in fact, are an important energy source because they are needed by spermatozoa for swimming [49], in eggs for larval development and survival [50].

In conclusion, we propose an artificial diet to define a system offshore of sea urchin culture, in order to obtain gonads with comparable quality to that of wild type. This method permits to preserve the wild stocks and satisfy the market demand.

Author details

Bernardetta Anna Tenuzzo[1], Elisabetta Carata[1], Stefania Mariano[1] and Luciana Dini[1,2]*

*Address all correspondence to: luciana.dini@unisalento.it

1 Department of Biological and Environmental Science and Technologies, University of Salento, Lecce, Italy

2 CNR Nanotec, Lecce, Italy

References

[1] Sellem F, Langar H, Pesando D. Age et croissance de l'oursin *Paracentrotus lividus* Lamarck, 1816 (*Echinodermata-Echinoidea*) dans le golfe de Tunis (Méditerranée). Oceanologica Acta. 2000;**23**(5):607-613. DOI: 10.1016/S0399-1784(00)0115-4

[2] Tenuzzo B, Zaccarelli N, Dini L. The reproductive cycle of the commercial sea urchin *Paracentrotus lividus* (Lamark, 1816; *Echinodermata-Echinoidea*) in the Ionian Sea. Italian Journal Zoology. 2012;**79**(2):200-211

[3] Diele K, Koch V, Saint-Paul U. Population structure, catch composition and CPUE of the artisanally harvested mangrove crab *Ucidescordatus* (*Ocypodidae*) in the Caetèestary, North Brazil: Indications for overfishing?. Aquatic Living Resources. 2005;**18**:169-178

[4] Pais A, Chessa L, Serra S, Ruiu A, Meloni G, Donno Y. The impact of commercial and recreational harvesting for *Paracentrotus lividus* on shallow rocky reef sea urchin communities in north-western Sardinia, Italy. Estuarine, Coastal and Shelf Science. 2007;**73**:589-597

[5] Gianguzza P, Chiantore M, Bonaviri C, Cattaneo-Vietti R, Vielmini I, Riggio S. The effects of recreational *Paracentrotus lividus* fishing on distribution patterns of sea urchins at Ustica Island MPA (Western Mediterranean, Italy). Fisheries Research. 2006;**81**:37-44

[6] Liu H, Kelly MS, Cook EJ, Black KD, Orr H, Zhu JX, Dong SL. The effect of diet type on growth and fatty acid composition of sea urchin larvae, I. *Paracentrotus lividus* (Lamarck, 1816) (*Echinodermata*). Aquaculture. 2007;**265**:247-262

[7] Liu H, Kelly MS, Cook EJ, Black KD, Orr H, Zhu JX, Dong SL. The effect of diet type on growth and fatty acid composition of sea urchin larvae, II. *Psammechinus miliaris* (Gmelin). Aquaculture. 2007;**264**:263-278

[8] Vadas RL, Beal B, Dowling T, Fegley JC. Experimental field tests of natural algal diets on gonad index and quality in the green sea urchin, *Strongylocentrotus droebachiensis*: A case for rapid summer production in post-spawned animals. Aquaculture. 2000;**182**:115-135

[9] Robinson SM, Colborne L. Enhancing roe of the green sea urchin using an artificial food source. Bulletin of the Aquaculture Association of Canada. 1997;**97**:14-20

[10] Mendes C, Becarra R. Cultivation of Loxechinus albus in lantern nets. In: Lawrence JM, Guzman O, editors. Sea Urchin Fisheries and Ecology. Lancaster USA: DESTECH Publications; 2004. 374

[11] James, PJ. A comparison of roe enhancement of the sea urchin *Evechinus chloroticus* in sea-based and land-based cages. Aquaculture. 2006;**253**:290-300

[12] Kelly MS, Brodie CC, McKenzie JD. Somatic and gonadal growth of the sea urchin *Psammechinus miliaris* (Gmelin) maintained in polyculture with the Atlantic salmon. Journal of Shellfish Research. 1998;**17**:1557-1562

[13] Cook EJ, Kelly MS. Enhanced production of the sea urchin *Paracentrotus lividus* in integrated open-water cultivation with Atlantic salmon Salmo salar. Aquaculture. 2007;**273**:573-585

[14] Cook EJ, Kelly MS. Effect of variation in the protein value of the red macroalga *Palmaria palmata* on the feeding and growth of the sea urchins *Psammechinus miliaris* and *Paracentrotus lividus* (*Echinodermata*). Aquaculture. 2007;**270**:207-217

[15] Grosjean P, Spirlet C, Gosselin P, Vaitilingon D, Jangoux M. Land-based, closed-cycle echiniculture of *Paracentrotus lividus* (Lamark) (*Echinoidea*: *Echinodermata*): A long-term experiment at a pilot scale. Journal of Shellfish Research. 1998;**17**(5):1523-1531

[16] Shpigel M, McBride SC, Marciano S, Lupatsch I. Propagation of the European sea urchin *Paracentrotus lividus* in Israel. In: Lawrence JM, Guzman O, editors. Sea Urchins: Fisheries and Aquaculture. Pennsylvania: DESTECH Publications; 2004. p. 386

[17] Cook EJ, Kelly MS. Co-culture of the sea urchin *Paracentrotus lividus* and the edible mussel *Mytilus edulis l.* on the west coast of scotland, united kingdom. Journal of Shellfish Research. 2009;**28**:553-559

[18] Fernandez C, Boudouresque CF. Nutrition of the sea urchin *Paracentrotus lividus* (*Echinodermata*: *Echinoidea*) fed different artificial food. Marine Ecology Progress Series. 2000;**204**:131-141

[19] Boudouresque CF, Verlaque M. Ecology of *Paracentrotus lividus*. In: Lawrence JM, editor. Edible Sea Urchins: Biology and Ecology. Amsterdam: Elsevier; 2007. pp. 243-285

[20] Schlosser SC, Lupatsch I, Lawrence JM, Lawrence AL, Shpigel M. Protein and energy digestibility and gonad development of the European sea urchin *Paracentrotus lividus* (Lamarck) fed algal and prepared diets during spring and fall. Aquaculture Research. 2005;**36**:972-982

[21] Meidel SK, Scheibling RE. Effects of food type and ration on reproductive maturation and growth of the sea urchin *Strongylocentrotus droebachiensis*. Marine Biology. 1999;**134**:155-166

[22] Otero-Villanueva MDM, Kelly MS, Burnell G. How prey size, type and abundance affects foraging decisions by the regular echinoid *Psammechinus miliaris*. Journal of the Marine Biological Association of the UK. 2006;**86**:773-781

[23] Haya D, Régis MB. Comportement trophiqué de *Paracentrotus lividus* (Lam.) (*Echinodermata*: *Echinoidea*) soumis à six regimés alimentaires dans des conditions éxperimentales. Mesogée. 1995;**54**:35-42

[24] Chopin T, Bushmann AH, Halling C, Troell M, Kautsky N, Neori A. Integrating seaweeds into marine aquaculture systems: A key toward sustainability. Journal of Phycology. 2001;**37**:975-986

[25] Naylor RL, Williams SL, Strong DR. Aquaculture—A gateway for exotic species. Science. 2001;**294**:1655-1656

[26] Black KD, Pearson TH. The environmental impacts of marine fish cage culture. In: Black KD, editor. Environmental Impacts of Aquaculture. Boca Raton, FL: CRC Press; 2001. pp. 1-31

[27] Buschmann AH, Mora OA, Gòmez P, Böttger M, Buitano S, Retamales G, Vergara PA, Gutierrez A. *Gracilaria chilensis* outdoor tank cultivation in Chile: Use of land-based salmon culture effluents. Aquacultural Engineering. 1994;**13**:283-300

[28] Neori A, Shpigel M, Ben-Ezra D. A sustainable integrated system for culture of fish, seaweed and abalone. Aquaculture. 2000;**186**:279-291

[29] Hernàndez I, Martìnez-Aragòn JF, Tovar A, Pèrez-Llorèns JL, Vergara JJ. Biofiltering efficiency in removal of dissolved nutrients by three species of estuarine macroalgae cultivated with sea bass (*Dicentrarchus labbrax*) waste water 2. Ammonium. Journal of Applied Phycology. 2002;**14**:375-384

[30] Nelson SG, Glenn EP, Conn J, Moore D, Walsh T, Akutagawa M. Cultivation of *Gracilaria parvispora* (*Rhodophyta*) in shrimp-farm effluent ditches and floating cages in Hawaii: A two-phase polyculture system. Aquaculture. 2001;**193**:239-248

[31] Jones AB, Dennison WC, Preston NP. Integrated treatment of shrimp effluent by sedimentation, oyster filtration and macroalgal absorption: A laboratory scale study. Aquaculture. 2001;**193**:155-178

[32] Jones AB, Preston NP, Dennison WC. The efficiency and condition of oysters and macroalgae used as biological filters of shrimp pond effluent. Aquaculture Research. 2002;**33**: 1-19

[33] Luning K. SEAPURA: Seaweeds purifying effluents from fish farms: An EU project coordinated by the Wadden Sea Station Sylt. Wadden Sea Newsletter. 2001;**2001**:20-21

[34] Walker C, Unuma T, McGinn N, Harrington L, Lesser M. Reproduction of sea urchins. In: John Miller Lawrence Editor. Edible Sea Urchins: Biology and Ecology; 2001. pp. 5-26

[35] Unuma T, Yamamoto T, Akiyama T, Shiraishi M, Ohta H. Quantitative changes in yolk protein and other components in the ovary and testis of the sea urchin *Pseudocentrotus depressus*. Journal of Experimental Biology. 2003;**206**:365-372

[36] Leighton P. Contributions towards the development of echinoculture in North Atlantic waters with particular reference to *Paracentrotus lividus*. [doctoral dissertation]. National University of Ireland; 1995. p. 222

[37] Lawrence JM. Edible Sea Urchins: Biology and Ecology. Boston, USA: Elsevier; 2007

[38] Shpigel M, McBride SC, Marciano S, Ron S, Ben-Amotz A. Improving gonad colour and somatic index in the European sea urchin *Paracentrotus lividus*. Aquaculture. 2005;**245**(1-4):101-109

[39] Dincer T, Cakli S. Chemical composition and biometrical measurements of the Turkish sea urchin (*Paracentrotus lividus*, Lamark, 1816). Critical Reviews in Food Science and Nutrition. 2007;**47**(1):21-26

[40] Fernandez C, Pergent G. Effect of different formulated diets and rearing conditions on growth parameters in the sea urchin *Paracentrotus lividus*. Journal of Shellfish Research. 1998;**17**:1571-1581

[41] Hammer H, Hammer B, Watts S, Lawrence A, Lawrence J. The effect of dietary protein and carbohydrate concentration on the biochemical composition and gametogenic condition of the sea urchin *Lytechinus variegates*. Journal of Experimental Marine Biology and Ecology. 2006;**334**:109-121

[42] Fernàndez C. Effect of diet on the biochemical composition of *Paracentrotus lividus* (*Echinodermata*: *Echinoidea*) under natural and rearing conditions (effect of diet on biochemical composition of urchin). Comparative Biochemistry and Physiology Part A: Physiology. 1997;**118**(4):1377-1384

[43] Liyana-Pathirana CF, Shahidi Whittick A. The effect of an artificial diet on the biochemical composition of the gonads of the sea urchin (*Strongycentrotus droebachiensis*). Food Chemistry. 2002;**79**:461-472

[44] Angioni A, Addis P. Characterization of the lipid fraction of wild sea urchin from the Sardinian Sea (western Mediterranean). Journal of Food Science. 2014;**79**(2):C155-C162

[45] Pulz O, Gross W. Valuable products from biotechnology of microalgae. Applied Microbiology and Biotechnology. 2004;**65**(6):635-648

[46] Pacetti D, Alberti F, Boselli E, Frega NG. Characterisation of furan fatty acids in adriatic fish. Food Chemistry. 2010;**122**:209-215

[47] Spiteller G. Furan fatty acids: Occurrence, synthesis, and reactions. Are furan fatty acids responsible for the cardio protective effects of a fish diet?. Lipids. 2005;**40**:755-771

[48] Marsh AG, Watts SA. Biochemical and energy requirements of gonad development. In: Lawrence JM, editor. Developments in Aquaculture and Fisheries Science, Edible Sea Urchins: Biology and Ecology. Vol. 32; 2007. pp. 35-53

[49] Mita M, Nakamura M. Energy metabolism of sea urchin spermatozoa: the endogenous substrate and ultrastructural correlates. In: Jangoux M, Lawrence JM, editor. Echinoderm Studies; 2001. pp. 85-110

[50] Sewell MA. Utilization of lipids during early development of the sea urchin *Evechinus chloroticus*. Marine Ecology Progress Series. 2005;**304**:133-142

[51] Lozano J, Galera J, Lòpez S, Turon X, Palacìn C, Morera G. Biological cycles and recruitment of *Paracentrotus lividus* (*Echinodermata*: *Echinoidea*) in two contrasting habitats. Marine Ecology Progress Series. 1995;**122**:179-191

[52] Sanchez-Espana AI, Martinéz-Pita I, Garcia FJ. Gonadal growth and reproduction in the commercial sea urchin *Paracentrotus lividus* (Lamarck, 1816) (*Echinodermata*: *Echinoidea*) from southern Spain. Hydrobiologia. 2004;**5191-3**:61-72

[53] Kempf M. Recherchs d'ècologie comparée sur *Paracentrotus lividus* (Lmk.) et *Arbacia lixula* (L.). Recueil des Travaux de la Station Marine d'Endoume Faculte des Sciences de Marseille. 1962;**25**:47-116

[54] Antongiovanni M, Lercker G, Secchiari P. Metodi di analisi per lo studio della frazione lipidica del latte, dei prodotti di origine animale e degli alimenti zootecnici. Edizioni PLUS-Pisa University Press, Italy; 2007

[55] Santercole V, Delmonte P, Kramer JK. Comparison of separations of fatty acids from fish products using a 30-m Supelcowax-10 and a 100-m SP-2560. Lipids. 2012;**47**(3):329-344

[56] Bradford MM. A rapid and sensitive method for the quantitation of microgram quantities of protein utilizing the principle of protein-dye binding. Analytical Biochemistry. 1976;**72**:248-254

[57] Koehler LM. Differentiation of carbohydrates by anthrone reaction rate and color intensity. Analytical Chemistry. 1952;**24**:1576-1579

[58] Gianguzza P, Badalamenti F, Bonaviri C, Riggio S. The operational sex ratio of the sea urchins *Paracentrotus lividus* population: The case of the Mediterranean marine protected area of Ustica Island (Tyrrhenian Sea Italy). Marine Ecology. 2009;**30**:125-132

[59] Repolho T, Helena MH, Orlando L, Gago J. Brood-stock diet effect on sea urchin *Paracentrotus lividus* (lamarck,1816) endotrophic larvae development: Potential for their year-round use in environmental toxicology assessment. Ecotoxicology and Environmental Safety. 2011;**74**:584-592

[60] Sartori D, Gaioni A. Can sea urchins benefit from an artificial diet? Physiological and histological assessment for echinoculture feasibility evaluation. Aquaculture Nutrition. 2016;**22**:1214-1221

Chapter 5

Importance of Gamete Quality in Ecotoxicological Application: Natural versus Bred Population in *Paracentrotus lividus*

Sartori Davide, Lera Samantha, Silvia Giuliani, Simona Macchia, Lorenzo Morroni, David Pellegrini and Andrea Gaion

Additional information is available at the end of the chapter

http://dx.doi.org/10.5772/intechopen.69235

Abstract

Several approaches have been tested to respond to the depletion of wild stocks, from the production of seeds to the setting up of closed echinoculture systems, starting with fertilization of eggs with the consequent development to adult sea urchins. Hence, in the last years, our research group has focused on the assessment of a feasible and sustainable strategy aimed to ensure a rapid and effective gonadal growth of healthy gametes in recirculating aquaculture system (RAS) to employ in ecotoxicological application. In order to compare the health of obtained gametes with wild populations, the effectiveness of diets was evaluated with different biological parameters, such as fertilization and embryo-development test, and with histological analysis of gonads to appraise the stage of maturation. Moreover, the information regarding different breeding conditions of adults and genetic variability should be combined with the analysis of larval settlement and its requirements, demonstrating the importance of these parameters for the possible closure of the echinoculture cycle in RAS. Results achieved so far in terms of gonadal development and health of gametes have provided evidence of success in overcoming natural gaps between reproductive events in natural populations and an efficient and standardize breeding condition in RAS.

Keywords: *Paracentrotus lividus*, gonadal growth, sexual maturation, diets, recirculating aquaculture system

1. Introduction

The commercial fisheries and destructive harvesting methods employed to meet market demand, caused, in the last decades, a dramatic depletion of *Paracentrotus lividus* in Europe, especially in the Mediterranean, leading to a complete disappearance of urchins from areas of former abundance. This scenario was exacerbated by its worldwide recognition as one of the most reliable bioindicator species, and its gametes have been used for biological assays in marine pollution monitoring.

Several approaches have been tested to respond to the depletion of wild stocks, from the production of seeds (juveniles for out-planting) [1] to the induction of gonadal growth (GG) in organisms belonging to natural populations [2, 3], but the most challenging strategy could be the setting up of closed echinoculture systems, starting with fertilization of eggs with the consequent development to adult sea urchins [4, 5].

The reproductive cycle of echinoids has been extensively studied and documented since the early 1930 [6, 7] and several studies have been carried investigating different light/dark regimes, temperature, and supply of artificial diets to ensure a rapid gonadal growth and to promote an effective maturation of gametes of *P. lividus*.

The aim of the present chapter is to present the most effective breeding conditions to promote gonadal growth and sexual maturation in *P. lividus*. Particular emphasis will be put toward the work carried out by our research group in the assessment of a feasible and sustainable strategy aimed to ensure a rapid and effective gonadal growth of healthy gametes in recirculating aquaculture system (RAS) with pellets, macroalgae, and formulated diets rich in carotenoids (*Spinacia oleracea* and *Zea mays*), which have been demonstrated to be crucial for various biological function, especially during egg production and development [8–11].

In order to compare the health of obtained gametes with wild populations, the effectiveness of diets was evaluated with different biological parameters, such as fertilization and embryo-development test, and with histological analysis of gonads to appraise the stage of maturation.

The results obtained by our group demonstrated the good feasibility of a low-cost, easy to standardize and sustainable diet in rearing vulnerable species such as *P. lividus* in RAS.

2. Gonadosomatic growth and sexual maturation in *Paracentrotus lividus* natural population

Different factors can be significant for the somatic growth (SG) of *P. lividus*, such as water temperature, nutritional aspects of nutriment, and gonadal development [12]. Among these, previous research demonstrated that seasonal variations of growth rate seem to be predominantly related to water temperature with a dilatory effect in response to low temperatures; in fact, Le Gall [13] reported the absence of growth in the population of sea urchins in the English

Channel when temperature was between 4 and 7°C. From the literature on this topic, growth is noted to increase proportionally with increasing temperature between 7 and 18°C, with the optimum condition predominantly between 18 and 22°C above this condition, growth slows to a halt completely when temperature exceeds 28°C [14–17]. In the Mediterranean Sea, growth peaks during spring, when the water temperature is between 12 and 18°C, while sporadically occurs in autumn and is practically absent in winter [14–17]. As a result of these premises, it can be estimated that urchins of 2 cm in diameter are around 2 years old and it takes approximately 4–5 years for them to reach 4 cm in diameter [12, 16, 18–21].

P. lividus has an annual reproductive cycle. According to some authors, these species present a single spawning event [22, 23], while others support the hypothesis that in a year may occur two reproductive events [24–26]. The reproductive cycle of *P. lividus* has been studied in detail by several authors and is known as the cycle of many echinoids, which is influenced by various environmental factors such as temperature [22, 23], photoperiod [17, 22, 23], hydrodynamics conditions [27], and trophic availability [23, 25–27].

Gonadosomatic index (GSI) is defined as the ratio between the mass of gonads and the mass of the whole organisms [23]; generally, the highest GSIs are reached by bigger individuals, with size tests ranging from 40 to 70 mm, rather than individuals belonging to the 20–40 mm class size [28, 29]. Previous studies on the gonadal growth (GG) of *P. lividus* in the Mediterranean and in the Atlantic Ocean reported the presence of two growth peaks whose timing can vary considerably even in populations of neighboring adjacent areas: indicatively the first peak takes place in spring followed by the second in late-summer/autumn [22, 29, 30].

Both field and *in vitro* studies seem to confirm the correlation between SG and GG with food availability [21, 31] and high rate of organic matter ingestion [32]. Gonadal development in particular seems to be enhanced by temperatures comprised between 18 and 22°C [17], even if some contrasting data were obtained from *in situ* studies reporting developed gonads in well-fed subtidal populations both in open sea and in lagoon environments [12, 22, 33–35]. High GSIs have been measured in populations with either low or high densities: the former [36] as a consequence of a low competition for food, the latter where the substrate is populated by a few algal species [23]. On the basis of these information, it can be noted that individuals of *P. lividus* invest greatly in reproductive strategy in conditions of scarcity of food. Although, GG, could be supported by high supply of algal fragments or food of high nutritional value transported by current flow.

P. lividus has separate sexes and there is no sexual dimorphism, yet, hermaphroditism cases have been observed for this species [22, 37]. *In vitro*, sexual maturity is reached in individuals of size ranging between 13 and 20 mm and/or after 5 months [38, 39]; however, sexual maturity can be reached in natural populations over longer periods of time.

2.1. Gamete spawning in *Paracentrotus lividus*

According to Fenaux [25], although the production of gametes takes place up to a temperature of 8°C, spontaneous spawning is not possible under 13.5°C. Thus, the reproductive period at subtropical latitudes takes places, from autumn to spring, until temperatures remain below 20–22°C. Along the French Mediterranean coasts, two main reproductive events were observed: one between May and June and the other in September/October [25]. In contrast with this scenario, data in the literature report for Italian populations of *P. lividus* **in** one single spawning period from October to June [22, 23, 40].

Generally, during the spawning events, male and female of *P. lividus* aggregate simultaneously release their gametes in the water [41]. Although these episodes do not always involve all individuals of a population, the homogenized suspension of sperms and eggs can trigger and encourage the release of gametes by other sea urchins located in remote places [42]. In both cases of single or double spawning periods throughout the year, water temperature seems to play a key role in determining the start of the event. While two spawning events have been registered, the first episode occurs when the temperature reaches 14–16°C and the second when the temperature returns to these values after the summer [22, 25, 43, 44]. It has been hypothesized that the first release can also be prompted by lengthening of photoperiod (about 15 h of daylight) rather than the temperature, while the end of spawning events seems to be controlled exclusively by temperature [30, 45]. Even though the presence of one or two spawning periods can be observed within the same region among different habitats [46], according to Lozano et al. [23], the natural emission of gametes would occur exclusively during spring/early summer. It should be taken in consideration, though, that the presence of larvae and postmetamorphosis individuals (1 mm in diameter) in October would seem to reveal the presence of a spawning event in late summer. Boudouresque and Verlaque [47], in apparent contrast with the pace dictated by the above reported variables (water temperature, photoperiod, habitat), reported that spawning can occur almost year-round, although not profusely. This adaptation can be a strategy to facilitate the dispersal of larvae and ensure greater reproductive success of the species.

2.2. Getting started: An overview on formulated diets to enhance gonadal growth and sexual maturation

To date, several studies have been conducted to identify the most suitable breeding condition for *P. lividus*, and frequently the results shown in these studies are not in accordance. Diverse formulated diets are available to promote GG, some formulated diets of particular interest among them are reported in **Table 1**.

Basuyaux and Blin [54, 55] tested six different diets (three of which were algal-based and the other three being based on maize) to enhance SG of the sea urchin *P. lividus* in a semi-closed system; their study reported the best growth rates with a mixed diet (*maize* and *Palmaria palmata*) and the lowest results obtained with the administration of algal-based diets. These results find confirmation in the data described by Garmendia et al. [56] after 60 days of rearing with carrot and algae (*Gelidium sesquipedale*). Frantzis and **Grémare** [32] rearing over a 6-month period sea urchin with macrophytes achieved the highest ingestion and gonadal production with *Rissoella verruculosa*, which is frequently reported among the most preferred species of seaweed for *P. lividus* despite the presence of repellent and toxic metabolites such as brominated compounds in it [57, 58].

	Tested species	Composition	Reference
Texas A&M formulation	*Lytechinus variegatus*	Three different formulation based on protein content: final protein concentrations vary from 9 to 31% dry weight (as fed basis)	[48]
NIWA feed	*Evechinus chloroticus*	• Protein 40.8% by dw; • Carbohydrates 26.2% by dw	[49]
NIFA feed	*Evechinus chloroticus*	• Protein 24.5% by dw; • Carbohydrates 49.6% by dw	[49]
Lawrence diet	*Loxechinus albus, Strongylocentrotus franciscanus* and *Strongylocentrotus purpuratus*	• Total crude protein 12–24% by wt 15–20% by wt; • Total carbohydrate 30–60% by wt 40–55% by wt; • Total crude fat 3–9% by wt. 4–7% by wt	[50]
Micciche' et al.	*Paracentrotus lividus, Loxechinus albus, Strongylocentrotus granularis, Strongylocentrotus intermedius*	• Egg albumen 45–55% by wt; • *Lactuca sativa* 30–40% by wt; • Fish flour 4–6% by wt.	[51]
Ross Island Salmon Ltd. feed	***Strongylocentrotus droebachiensis***	• Rockweed meal *(Ascophyllum nodosum)* 22.84% by dw; • Soybean meal 27.92% by dw; • Dulse powder *(Palmaria palmata.)* 10.00% by dw; • Lecithin 1% by dw; • Canola oil 2% by dw	[52]
St. Andrews Biological Station feed	***Strongylocentrotus droebachiensis***	• Rockweed meal *(Ascophyllum nodosum)* 8.40% by wt; • Soybean meal 45.00% by wt; • Lecithin 2% by dw; • Canola oil 4% by wt.	[53]
Pliva-Kalinovica-Zagreb factory diet	***Paracentrotus lividus***	• Crude protein 12.6–22.4% by dw; • Fat 7.6% by dw; • Crude fiber 1.8–2.4%	[54, 55]

Table 1. Formulated diet of particular interest to promote gonadal growth with their major component.

Studying the effect of light regime and temperature in rearing condition, indications reported in the literature are once again not consistent: McCarron [59] showed that darkness supports higher SG than the photoperiod treatment whereas other works exhibited that the most suitable rearing condition to enhance gonadal growth were short photoperiod (9 h daylight) and temperature ranging between 18 and 24°C [17, 45]. In addition to these results, for a short rearing period, Fabbrocini et al. [60] found that nutrition appears to be the key factor in the induction of GG.

Comparing the results obtained with "vegetal" and "animal" diet, Fernandez and Boudouresque [61] suggested that the latter resulted in the highest absorption rate, and consequently the best GG; on the contrary, the highest ingestion rate was obtained with vegetable feed. With regard to absorption, results showed that it was negatively correlated with ingestion and carbohydrate level content of the food. These data were confirmed by further works and substantially the prepared diets, characterized by high protein content, guaranteed a better GG in comparison with vegetable diet or low protein content diet [62–65]. Luis and co-worker [66] studying the effect of plant-based diet on spawning performance of *P. lividus* in captivity recorded that a mixed diet maize and seaweed gave better results (79% of the tested urchins) if compared to pure maize diet (50%) and the pure seaweed diet (36%) in terms of consistency of spawning throughout the year.

In conclusion, when the reproductive conditions (organisms in mature or premature stage) allow short breading periods, the synchronization of the gonads to the emission stages can be induced simply by controlling the diet, without the need for altering the photoperiod and stressed the animals with unnatural photoperiod.

2.3. Breeding condition to promote sexual maturation in RAS

Interest in cultivation of sea urchins has increased over the last 2 decades as a consequence of the depletion of wild stocks [1, 22, 47, 62, 67, 68]. In this context, breeding of *P. lividus* for restocking wild populations in addition to its use for human consumption is an aspect to consider for future developments. An additional reason for that is since 1980, it was recognized worldwide among the most reliable bioindicator species [69], and its gametes have been extensively used for biological assays [70–89]. Therefore, the development of rearing techniques for this species is a current issue for both production of gametes for ecotoxicological application and restocking depleted natural populations due to the growing market demand of roe; a request that otherwise natural populations are unable to meet [83].

Hence, in the last years our research group has focused on the assessment of a feasible strategy aimed to ensure a rapid and effective GG of healthy gametes of *P. lividus* in RAS.

2.3.1. Maintenance of mature stage in Paracentrotus lividus in RAS

Our first goal was focused on the maintenance in RAS over a 4-month period of *P. lividus* specimens in mature stage collected during spawning period. After initial 5 days acclimatization fasting period, individuals were reared in the recirculating aquaculture system at 14°C temperature and photoperiod of 10 h L: 14 h D testing two different diets. The first diet was composed of 50% (in volume) maize kernel that had been previously crushed with a blender into grains of a few millimeters and the remaining 50% (in volume) by a mixture of fresh seaweed (*Dyctiopteris* sp., *Padina Pavonica, Dyctiota* sp., *Ulva lattuga, Halopteris scoparia, Flabellia petiolata, Laurencia* sp., *Corallina elongata, Codium* sp.), collected from the same sampling site of sea urchin. The second diet consisted of a mixture in equal proportion of maize and freshly chopped spinach leaves. Spinach, such as maize, is a vegetable with a high nutritional value and high content of carotenoids [83]. Several studies have demonstrated that spinach leaves contain natural antioxidants with potential biological activities [90–92]. Indeed polyphenols are now widely accepted as physiological antioxidants that have significant activities and capacity to protect critical macromolecules against the numerous degenerative diseases linked to free reactive oxygen species (ROS) [93–95].

Before being administered, the ingestion rates of every single element of diets were assessed in order to establish the most suitable feed for sea urchins in RAS (**Table 2**). Results showed that in order of preference, the most appreciated food were maize, *Dyctiopteris* sp., spinach, and *Codium* sp.

Considering that brown algae together with *Posidonia oceanica* leaves are among the main components of adult *P. lividus* diet [96–98], with regard to the brown alga *Dyctiopteris* sp. results confirmed this expectation. On the contrary, it is not as clear as the poor desirability for the other brown algae administered. The low appreciation of red algae, with the exception of *Laurencia* sp., could be related to the presence of brominate substances as reported by Codomier et al. [57] for the red alga *Asparagopsis armata*. Among the seaweed provided, *P. lividus* clearly did not show any preference toward algae that have a coriaceous consistency due to the presence of precipitates of calcium carbonate in the structure of the alga (*Corallina elongata, Padina pavonica,* and *Flabellia petiolate*). The poor desirability toward *Halopteris scoparia* could be due to the presence of phenolic compounds, which act as a deterrent for *P. lividus* [99]. However, in contrast with what reported in literature [47], *P. lividus* used in our experiments did not ingest important quantities of the green alga *Ulva lactuca* (ingestion rate [g/day + sd]: 0.030 ± 1.06 by dw, see **Table 2**). As already mentioned above, among the red algae, an appreciable consumption has been registered for *Laurencia obtusa*; these data seem to be in contrast with those reported in Boudouresque and Verlaque [47].

Species	Ingestion rates (g/day dw)
Codium sp.	0.137 ± 0.32
Corallina elongata	0.025 ± 0.68
Dyctiopteris sp.	0.141 ± 0.45
Dyctiota sp	0.076 ± 1.16
Flabellia petiolata	0.030 ± 2.08
Halopteris scoparia (=*Stypocaulon scoparium*)	0.035 ± 2.13
Laurencia sp.	0.115 ± 0.89
Zea mays (*crusched Kernel*)	0.281 ± 0.93
Padina pavonica	0.085 ± 0.56
Spinacia oleracea	0.139 ± 1.18
Ulva lactuca	0.030 ± 1.06

Table 2. Ingestion rates by for the macrophyte, maize, and spinach tested. Ingestion rates are expressed in terms of dry weight per day. The values are mean and standard deviation.

The consumption of spinach resulted to be comparable to the "most preferred algae"; however, the consumption of maize was significantly higher than those recorded for macrophytes. This unusual feeding behavior could be partly explained by the higher content of carbohydrates and proteins in maize with respect to algae and spinach. Indeed, these two ingredients exhibit a very similar biochemical composition, characterized by high water content and moderate presence of proteins and lipids. Maize is a primary source of energy supplement and can provide up to 30% protein, 60% energy, and 90% starch in animal diets [100] and is an important source of carotenoids. Echinenone and (60R)-β-carotene-4-one are the major carotenoids in both ovaries and testes [101, 102]; in gonads, echinenone accounts for approximately 50–60% of the total pigment [11, 103] and it is metabolized from dietary β-carotene. It is transported to or stored into gonads in much greater concentrations than other carotenoids, where it may play a role both in production of eggs, development, and immunological modulation [9, 102, 104, 105].

To evaluate the quality of gametes and embryos obtained from organisms maintained into mature stage in RAS, fertilization, and embryo-development tests by means of reference toxicant were performed every month during treatment with gametes collected by reared individuals.

Diets tested in these trials, in combination with a 10 h L:14 h D light regime and a water temperature of 14°C, have ensured the maintenance of animals into a mature prespawning stage for a 4-month period [83]. This result permitted to overcome the summer months during which, at our latitude, it is not possible to obtain gametes from organisms belonging to wild populations [40]. EC_{50} values obtained, both for fertilization and embryo-development tests during the experiments, were consistent with the laboratory control chart and those reported in the literature for different species of echinoderms, including *P. lividus* [10, 78, 82, 106–109]. It is worth reporting that the gametes obtained by tested individuals were comparable to those of natural populations in terms of response to the reference toxicant copper(II) nitrate ($Cu(NO_3)_2 \times 3H_2O$) [83].

2.3.2. Reliable breeding condition to promote sexual maturation in RAS

P. lividus has an annual reproductive cycle, although according to some authors these species presents a single spawning event [22, 23], whereas others support the hypothesis that two reproductive events may occur in a year [24–26]. The reproductive cycle of *P. lividus* has been well studied, and much research has been carried out to determine all its phases in relation to temperature [22, 23, 29, 30] photoperiod [17, 22, 23], hydrodynamics conditions [27], and trophic availability [23, 25–27].

After having identified the suitable conditions to maintain *P. lividus* in mature stage (stages III–V) [22] in RAS, our research group focused on the analysis of fast reliable breeding conditions to promote GG and sexual maturation in *P. lividus* in order to have a continuous production of gametes for scientific research and ecotoxicological tests. Different diet treatments to enhance gametes maturation were tested by our research group in a previous study [83], focusing in particular on the reliability of a maize and spinach diet (MSD), a macroalgae diet (MD), and a diet based on a commercial pellet normally used in aquaculture for warm-water species (Classic K®, PD). The biochemical composition of pellet Classic K® (Hendrix S.p.A) employed in our study is shown in **Table 3**.

Composition	%
Protein	43.0
Crude fat	11.5
Crude fiber	3.2
Ash	8.0
Phosphorus	0.8
Digestible energy (MJ/kg)	14.8

Table 3. Biochemical composition (%) of pellet Classic K® (Hendrix S.p.A.).

The MD and the MSD were the same, previously employed in the maintenance of mature stage in *P. lividus* reared in RAS.

Before being fed *ad libitum* with three chosen diets, sea urchins were starved for 2 months in order to promote the reabsorption of gonads and get them in phase regarding their reproductive cycle [30]. Specimens were kept in aquaria with 12 ± 1°C water temperature and exposed to a photoperiod 12 h L:12 h D completely deprived of food [4]. To promote the maturation of adult *P. lividus,* diets have been tested for 9 weeks in combination with a photoperiod of 10 h L:14 h D and water temperature of 16°C. In order to evaluate the effectiveness of the different diets, multiple biological data were analyzed, in particular: gonad index (GI), histological examination of gonadic tissue; analysis by using harmonic generation (HGM) and two-photon photon (2PF) microscopy; fertilization and embryo-development test with a reference toxicant ($Cu(NO_3)_2 \times 3H_2O$).

The second and third harmonic generation microscopy (SHG-THG) and the 2PF are nonlinear microscopy techniques, which base their optical resolution on the interaction of the wavelength of light with matter.

The THG and SHG techniques are a nonfluorescent multiphoton technique of laser scanning microscopy that allow to acquire signals with submicron spatial resolution without the use of fluorescent markers [111]. In copepods and zebrafish, these techniques have been shown to be able to reveal the onset of cell death mechanisms (apoptosis) [111–113]. In particular, the signal of the SHG can reveal the distribution of collagen fibers and striated muscle myosin [114], while THG microscopy is more versatile than SHG microscopy and can highlight, through the discontinuity of the refractive index, the morphology of cell membranes and of lipid vesicles [115]. The microscope used for this work is based on a femtosecond laser Cr-forsterite, which operates around 1230 nm. This laser is able to penetrate deep into the tissue causing little damage compared with the common Ti: sapphire laser used in fluorescence microscopy (700–1000 nm). The laser was mounted on an Olympus BX51 microscope, and plutei obtained from gametes of reared organisms in recirculating aquaculture system were observed with a 60× immersion objective and numerical aperture (NA) of 1.2 (**Figure 1**) at the Molecular Imaging Center, National Taiwan University, Taiwan.

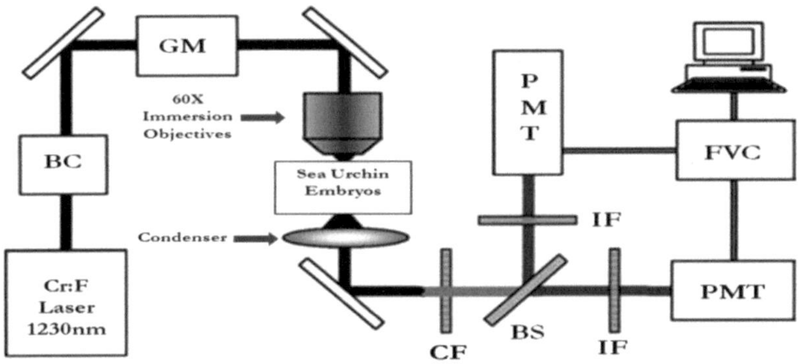

Figure 1. Schematic diagram of the microscope: BC, beam collimator; GM, galvanometric mirrors; CF, color filter; BS, beamsplitter; IF, interference filter; PMT, photomultiplier tube; FVC, fluoview control unit [110].

Tested diets were successful in stimulating gametogenesis and ensuring the production of healthy gametes in short time (3 weeks); among these, MSD gave the best results in terms of GI values. According to histological analysis, only MSD and PD diets were suitable to guarantee a rapid transition of sea urchins from an inactive phase (stage VI-Spent) to an active phase of gametogenesis (stages II–V) [22]. These results confirmed that photoperiod, temperature, and diet allowed to maintain the constant presence of mature *P. lividus* in the rearing tanks and overcome summer period during which sea urchins belonging to natural population are unable to produce gametes [40].

EC_{50} values obtained from fertilization and embryo-development tests with reference toxicant (**Figure 2a** and **b**) were consistent with those reported in the literature, demonstrating the good quality of gametes [10, 72, 82, 106, 109]. In particular, sperm obtained from animal reared with MD diet seems to be more sensitive toward copper(II) nitrate if compared with gametes obtained from other diets or belonging to natural population. This aspect reveals that the high energetic value of PD diet (protein = 43.0% and crude fat = 11.5%) and the richness in carotenoids and antioxidant in MSD diet proved to be important factors in the ability of sperm to cope with toxicity resulting from exposure to the reference metal [83]. The effect of the biological richness of MSD in terms of energy source, carotenoids, and antioxidants can be seen in the high EC_{50} value; indeed after a 9-week treatment, plutei demonstrated to have a resistance against copper(II) nitrate comparable with the wild population, higher than plutei bred with MD or PD diet (**Figure 2b**); these data were further confirmed by histological evidence [116].

Indeed, as previously demonstrated, thanks to the combination of the high nutritional value due to carbohydrates content and fairly good presence of protein in maize [117] with the remarkable richness in carotenoids and polyphenols of spinach [90, 118], MSD holds the essential chemical compounds to protect biological tissues against oxidative processes [119, 120].

Nevertheless, both field and *in vitro* studies seemed to confirm that somatic and GG ensued when food availability and organic matter ingested was high [21, 31, 32], although other physical parameters such as light regime and temperature can positively affect it [59].

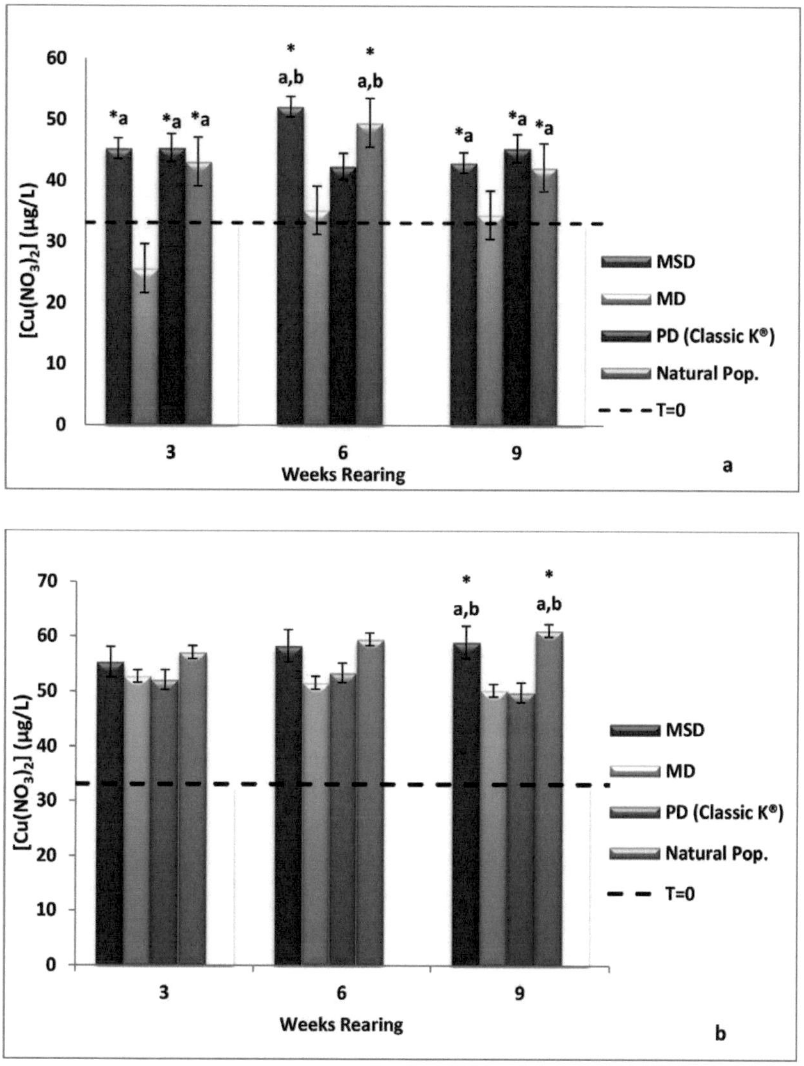

Figure 2. EC_{50} (µg/L) values of reference toxicant $[Cu(NO_3)_2 \times 3H_2O]$ obtained with fertilization (a) and embryo development (b) tests performed on *Paracentrotus lividus* reared in a recirculating aquaculture system (RAS) with artificial diets maize and spinach (MSD), pellet Classic K® (PD) and macroalgae (MD). The EC_{50} values obtained are compared with those obtained from *P. lividus* belonging to a natural population (Natural Pop.). EC_{50} (µg/L) values obtained at $T = 0$ for the wild population are reported. The values are mean and standard deviation [116]. "Note: *a= statistically sigfootnificant with respect to MD diet; *b statistically significant with respect to PD diet".

This chapter reviews innovative techniques established by our research group with regard to sea urchin plutei developed from gametes of reared organisms. Results concerning bred organisms were compared with those obtained from organisms belonging to the natural population in order to evaluate and assess possible events of cell death or abnormalities development biological damage linked with the rearing conditions and diets employed.

The evaluation of effects induced by rearing conditions by means of HGM microscopy techniques highlighted differences between plutei from the three diets. In particular, plutei obtained from the MSD diet presented an increase in fluorescence signal, both with 2PF and THG technique that we could hypothesize be related to apoptotic [121] or autophagy event [122, 123] (**Figure 3**). On the contrary, natural larvae or larvae obtained after PD or MD diets did not show any increase in fluorescence signal. Another factor to consider when evaluating MSD diet is that plutei showed a loss of THG signal in the skeleton, potentially due to incorrect tissue organization in the skeletal structures, phenomenon previously observed in *P. lividus* plutei exposed to $HgCl_2$ [121]. Considering these data, although feasible both for GG and gametes maturation in short rearing period, MSD could lead to the generation of plutei unstable for ecotoxicological and echinoculture application. Further investigations are needed with regard to this topic, considering that programmed cell death is a physiological process aimed to prepare the tissues of *P. lividus* larvae before metamorphosis [124].

The HGM microscopy technique applied in this study allowed the observation of abnormalities in the development of sea urchin plutei obtained from *P. lividus* kept in RAS with different diets. This technique is certainly a valuable and promising tool for applications in ecotoxicological studies, as confirmed by other *in vivo* studies with other model organisms such as zebrafish embryos (*Danio rerio*) [111, 112]. The same apoptotic body in the zebrafish hindbrain was positively stained through the fluorescent marker acridine orange. In a study conducted on nauplii of *Acartia tonsa* authors reported that the strong fluorescent signal detected with the 2PF and with the THG was associated with the onset of apoptosis in the digestive system of copepods, data confirmed by the classical staining technique of TUNEL

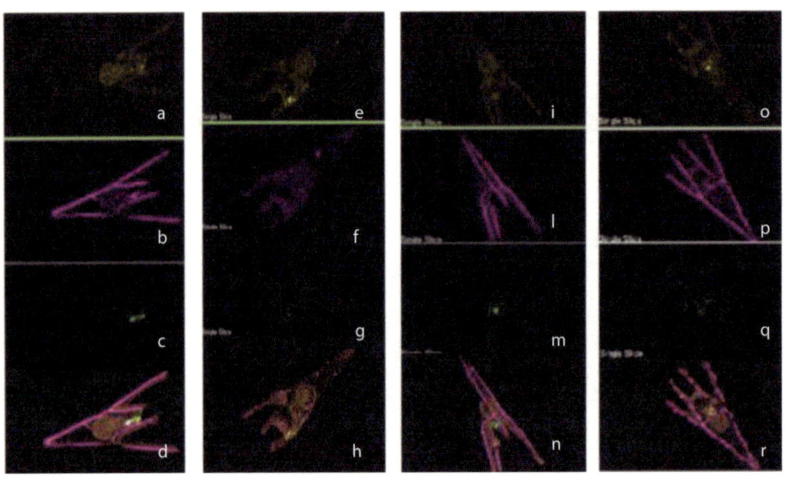

Figure 3. Sea urchin plutei obtained from different diets observed with light microscopy, two photon fluorescence (2PF) microscopy (a, e, i, o), third (THG) (b, f, l, p), and second (SHG) harmonic generation microscopy(c, g, m, q). The images obtained merging the THG, 2PF, and SHG signals are presented in images d, h, n, r. Plutei observed are obtained from *Paracentrotus lividus* belonging to natural population (a–d) and reared with maize and spinach (e–h), pellet Classic K® (i–n) and macrophytes (o–r).

[113]. In our experience, the noninvasive nature of the SHG and THG technique has permitted a three-dimensional observation of the cellular structures of sea urchin pluteus allowing the observation of morphological changes, in the complex development processes related to the rearing condition. In addition, these techniques provided important results without the use of fluorescence markers, overcoming the common phenomena of photodamage, phototoxicity, and photobleaching linked to the use of fluorescent probes.

3. Further development

Results achieved so far in terms of gonadal development and health of gametes have provided evidence of success in overcoming natural gaps between reproductive events in natural populations and an efficient and standardize breeding condition in RAS. In the past few years, different authors published results originating from different cultivating conditions, considering various environmental parameters as well as diets. Raposo et al. [125], for example, tested the efficacy of three artificial diets (A, green macoalgae *Codium* sp.; B, solid mix diet, with macroalgae *Codium* sp., carrot and cabbage; and C, maize and spinach) during 80 days to promote gonadal growth and the maturation of *P. lividus* gametes. In their experiments, a temperature of 20 ± 1°C was set during the dietary treatment, whereas in our trials this temperature caused spawning in specimens compromising the continuous collection of gametes. Sanja et al. [126] tested the efficacy of four different diets in a semiclosed recirculating system. In this experiment, temperature and salinity of seawater were not kept constant but allowed to vary according to natural environment; in particular, they ranged between 15 and 23°C and 32–37‰, respectively. Colak et al. [127] opted for a temperature of 22°C and a salinity of 38‰ and a diet based on pellet to cultivate urchins for histological analysis. The variability of results combined by the different environmental conditions of the tanks suggests that different populations of *P. lividus*, sampled by diverse areas and in different periods of the year, can respond in dissimilar ways to aquaculture conditions. To confirm this hypothesis, we compared the toxicological responses of three different populations toward three contaminants; these populations where distributed in a radius of ca. 10 km and the sampling activity was performed in five different times of the year [128]. Results showed a high variability of responses, even if considering a small-scale variation of populations. This result suggests that further analyses are required in order to assess the variability of a larger distribution of populations and the contribution of the genetic variability. Therefore, genetic analyses will be performed on the urchins belonging to these populations, with the expected results useful to comprehend the genetic variability and its correlation with the results previously obtained.

Finally, the information regarding different breeding conditions of adults and genetic variability should be combined with the analysis of larval settlement and its requirements. Indeed, Colin et al. [129] demonstrated that settlement rates of competent urchin larvae were significantly correlated with different substrates. Brundu et al. [130] combined the analysis of two different settlement substrates with four larval dietary treatments on the survival and growth of the larvae, demonstrating the importance of these parameters for the possible closure of the echinoculture cycle in RAS.

Author details

Sartori Davide[1]*, Lera Samantha[2], Silvia Giuliani[1], Simona Macchia[1], Lorenzo Morroni[1], David Pellegrini[1] and Andrea Gaion[3]

*Address all correspondence to: davide.sartori@isprambiente.it

1 Italian National Institute for Environmental Protection and Research—ISPRA, Piazzale dei Marmi, Livorno, Italy

2 Agenzia Regionale per la Protezione Ambientale della Toscana—ARPAT, Lucca, Italy

3 South Devon College, Vantage Point, Paignton, UK

References

[1] Yokota Y. Fishery and consumption of the sea urchin in Japan. In: Yokota Y, Matranga V, Smolenicka Z, editors. The Sea Urchin: From Basic Biology to Aquaculture. Lisse, The Netherlands: Swets & Zeitlinger BV; 2002. pp. 129-138

[2] Cook EJ, Kelly MS. Enhanced production of the sea urchin *Paracentrotus lividus* in integrated open-water cultivation with Atlantic salmon *Salmo salar*. Aquaculture. 2007; **273**:573-585

[3] Pantazis PA. The culture potential of *Paracentrotus lividus* (Lamarck 1816) in Greece: A preliminary report. Aquaculture International. 2009;**17**:545-552

[4] Grosjean P, Spirlet C, Gosselin P, Vaitilingon D, Jangoux M. Land-based, closed-cycle echiniculture of *Paracentrotus lividus* (Lamarck) (Echinodermata: Echinoidea): A long-term experiment at a pilot scale. Journal of Shellfish Research. 1998;**17**:1523-1531

[5] Devin MG. Land-based echinoculture: A novel system to culture adult sea urchins. In: Yokota Y, Matranga V, Smolenicka Z, editors. The Sea Urchin: From Basic Biology to Aquaculture. Lisse, the Netherlands: Swets & Zeitlinger BV; 2002. pp. 145-159

[6] Moore HB. A comparison of the biology of *Echinus esculentus* in different habitats. Part I. Journal of the Marine Biological Association of the UK. 1934;**19**:869-881

[7] Boolootian RA. Reproductive physiology. In: Boolotian RA, editor. Physiology of Echinodermata. New York, USA: Interscience Press; 1996. p. 822

[8] Matsuno T. Xanthophylls as precursors of retinoids. Pure and Applied Chemistry. 1991; **63**:81-88

[9] Bendich A. Recent advances in clinical research involving carotenoids. Pure and Applied Chemistry. 1994;**66**:1017-1024

[10] Tsushima M, Kawakami T, Mine M, Matsuno T. The role of carotenoids in development in sea urchin *Pseudocentrotus depressus*. Invertebrate Reproduction and Development. 1997;**32**(2):149-153

[11] Kawakami T, Tsushima M, Katabami Y, Mine M, Ishida A, Matsuno T. Effect of β,β-carotene, β-echinenone, astaxanthin, fucoxanthin, vitamin A and vitamin E on the biological defense of the sea urchin *Pseudocentrotus depressus*. Journal of Experimental Marine Biology and Ecology. 1998;**226**:165-174

[12] Fernandez C. Croissance et nutrition de *Paracentrotus lividus* dans le cadre d'un projet aquacole avec alimentation artificielle [Thesis]. Corte: Université de Corse; 1996

[13] Le Gall P. Culture of echinoderms. In: Barnabé G, editor. Aquaculture. Vol. 1. New York: Ellis Horwood; 1990. pp. 443-462. DOI: 10.1017/S0025315400037553

[14] Azzolina JF. Contribution à l'étude de la dynamique des populations de l'oursin comestibles *Paracentrotus lividus*. Thèse Doct, Univ Aix- Marseille 2. 1988

[15] Fernandez C, Caltagirone A. Growth rate of adult sea urchins, *Paracentrotus lividus* in a lagoon environment: The effect of different diet types. In: David B, Guille A, Féral JP, Roux M, editors. Echinoderms Through Time. Rotterdam: AA Balkema Press; 1994. pp. 655-660

[16] Turon X, Giribert G, López S, Palacín C. Growth and population structure of *Paracentrotus lividus* (Echinodermata: Echinoidea) in two contrasting habitats. Marine Ecology Progress Series. 1995;**122**:193-204

[17] Shpigel M, McBride S, Marciano S, Lupatsch I. The effect of photoperiod and temperature on the reproduction of European sea urchin *Paracentrotus lividus*. Aquaculture. 2004;**232**:343-355

[18] Grosjean P, Spirlet C, Jangoux M. Experimental study of growth in the echinoid *Paracentrotus lividus* (Lamarck, 1816) (Echinodermata). Journal of Experimental Marine Biology and Ecology. 1996;**201**(1-2):173-184

[19] Sellem F, Langar H, Pesando D. Age et croissance de l'oursin *Paracentrotus lividus* Lamarck, 1816 (Echinodermata-Echinoidea) dans le golfe de Tunis (Méditerranée). Oceanol Acta. 2000;**23**(5):607-613

[20] Gago J, Range P, Luis O. Growth, reproductive biology and habitat selection of the sea urchin *Paracentrotus lividus* in the coastal waters of Cascais, Portugal. In: Féral JP, David B, editors. Echinoderm Research 2001. Lisse: AA Balkema Press; 2003. pp. 269-276

[21] Grosjean P, Spirlet C, Jangoux M. A functional growth model with intraspecific competition applied to a sea urchin, *Paracentrotus lividus*. Canadian Journal of Fisheries and Aquatic Sciences. 2003;**60**(3):237-246

[22] Byrne M. Annual reproductive cycles of the commercial sea urchin *Paracentrotus lividus* from an exposed intertidal and a sheltered subtidal habitat on the west coast of Ireland. Marine Biology. 1990;**104**:275-289

[23] Lozano J, Galera J, Lopez S, Turon X, Palacin C, Morera G. Biological cycles and recruitment of *Paracentrotus lividus* (Echinodermata: Echinoidea) in two contrasting habitats. Marine Ecology Progress Series. 1995;**122**:179-191

[24] Crapp GB, Willis ME. Age determination in the sea urchin *Paracentrotus lividus* (Lamarck) with notes on the reproduction cycle. Journal of Experimental Marine Biology and Ecology. 1975;**20**:157-178

[25] Fenaux L. Maturation des gonade set cycle saisonnier des larves chez A. lixula, *P. lividus* et *P. microturberculatus* (echinides) à Villefranche-Sur-Mer. Vie Milieu.1968;**3**:1-52

[26] Régis MB. Analyse des fluctuation des indices physiologiques chez deux échinoides (*Paracentrotus lividus* (Lmck) et *Arbacia lixula* L.) du Golfe de Marseille. Téthys. 1979; **9**:167-181

[27] Guettaf M, San Martin GA, Francour P. Interpopulation variability of the reproductive cycle of *Paracentrotus lividus* (Echinodermata: Echinoidea) in the south-western Mediterranean. Journal of the Marine Biological Association of the UK. 2000;**80**:899-907

[28] Martínez I, García FJ, Sànchez AI, Daza JL, del Castillo F. Biometric parameters and reproductive cycle of *Paracentrotus lividus* (Lamarck) in three habitats of the southern Spain. In: Féral JP, David D, editors. Echinoderm Research 2001. Lisse: AA Balkema Press; 2003. pp. 281-287

[29] Sánchez-España AI, Martínez-Pita I, García FJ. Gonadal growth and reproduction in the commercial sea urchin *Paracentrotus lividus* (Lamarck, 1816) (Echinodermata: Echinoidea) from southern Spain. Hydrobiologia. 2004;**519**:61-72

[30] Spirlet C, Grosjean PH, Jangoux M. Reproductive cycle of the echinoid *Paracentrotus lividus*: Analysis by means of the maturity index. Invertebrate Reproduction and Development. 1998;**34**:69-81

[31] Lawrence J, Fenaux L, Corre MC, Lawrence A. The effect of quantity and quality of prepared diets on production in *Paracentrotus lividus* (Echinodermata: Echinoidea). In: Scalera-Liaci L, Canicatti C, editors. Echinoderm Research 1991. Rotterdam: AA Balkema Press; 1992. pp. 107-110

[32] Frantzis A, Grémare A. Ingestion, absorption, growth rate of *Paracentrotus lividus* (Echinodermata: Echinoïdea) fed different macrophytes. Marine Ecology Progress Series. 1992;**95**:169-183

[33] Fernandez C. Recherches preliminaries à la mise en placed'un ppilote d'aquaculture de l'oursin *Paracentrotus lividus*. Mém Maîtrise Sci Techn, Univ Corse. 1990

[34] San Martin G. Contribution à la gestion des stocks d'oursins: Etude des populations et transplantations de *Paracentrotus lividus* à Marseille (France, Méditerranée) et production de *Loxechinus albus* à Chiloe (Chili, Pacifique) [Thesis]. Université d'Aix – Marseille; 1995

[35] Fernandez C, Boudouresque CF. Phenotypic plasticity of *Paracentrotus lividus* (Echinodermata: Echinoidea) in a lagoonal environment. Marine Ecology Progress Series. 1997; **152**(1-3):145-154

[36] Guettaf M, San Martin GA. Etude de la variabilité de l'indice gonadique de l'oursin comestible *Paracentrotus lividus* (Echinodermata: Echinidae) en Méditerranée nord-occidentale. Vie Milieu. 1995;**45**(2):129-137

[37] Neefs Y. Sur divers cas d'hermaphrodisme fonctionnel chez l'oursin *Strongylocentrotus lividus*. CR Acad Sci Paris. 1937;**204**(11):901-902

[38] Azzolina JF. Evolution à long terme des populations de l'oursin comestible *Paracentrotus lividus* dans la baie de Port-Cros (Var, France). In: Boudouresque CF, editor. Colloque international sur *Paracentrotus lividus* et les oursins comestibles. Marseilles: GIS Posidonie; 1987. pp. 257-269

[39] Cellario C, Fenaux L. *Paracentrotus lividus* (Lamarck) in culture (larval and benthic phases): Parameters of growth observed two years following metamorphosis. Aquaculture. 1990;**84**:173-188

[40] Giambartolomei FM. Nuove acquisizioni sul test di fertilità dello sperma di riccio di mare *Paracentrotus lividus* [Thesis] Italy: University of Padova; 1990

[41] Cherbonnier G. Le roman des Echinodermes. Rennes: Les beaux Livres; 1954

[42] Kečkeš S, Ozretić B, Lucu Č. About a possible mechanism involved in the shedding of sea-urchins. Experentia. 1966;**22**:146-147

[43] Pedrotti ML. Spatial and temporal distribution and recruitment of echinoderm larvae in the Ligurian Sea. Journal of the Marine Biological Association of the UK. 1993;**73**:513-530

[44] Bayed A, Quiniou F, Benrha A, Guillou M. The *Paracentrotus lividus* population from the northern Moroccan Atlantic coast: Growth, reproduction and health condition. Journal of the Marine Biological Association of the UK. 2005;**85**:999-1007

[45] Spirlet C, Grosjean Ph, Jangoux M. Optimisation of gonad growth by manipulation of temperature and photoperiod in cultivated sea urchins, *Paracentrotus lividus* (Lamarck) (Echinodermata). Aquaculture. 2000;**185**:85-99

[46] Guettaf M. Contribution a l'étude de la variabilité du cycle reproductif (indice gonadique et histologie des gonades) chez *Paracentrotus lividus* (Echinodermata: Echinoidea) en Méditerranée sud occidentale (Algérie) [These]. Universite de la Mediterranee Aix-Marseille III; 1997

[47] Boudouresque CF, Verlaque M. Ecology of *Paracentrotus lividus*. In: Lawrence J, editor. Edible Sea Urchins: Biology and Ecology. 2nd ed. Amsterdam: Elsevier Press; 2007. pp. 243-285

[48] Woods CM, James P, Moss G, Wright J, Siikavuopio SI. A comparison of the effect of urchin size and diet on gonad yield and quality in the sea urchin Evechinus chloroticus Valenciennes. Aquaculture International. 2008;**16**:49-68

[49] Hammer H, Watts S, Lawrence A, Lawrence J, Desmond R. (2006) The effect of dietary protein onconsumption, survival, growth and production of the sea urchin, Lytechinus variegatus. Aquaculture. 2006;**254**:483-495

[50] Lawrence AL, Lawrence JM, Kearns JP, Rokey GJ. Sea urchin feed and method of producing same. United States Patent Number: 5637333. 1997

[51] Micciche' L, Mazzola A, Vizzini S, Scariano P, Falcone A. High-performance feed stuff formulation for aquaculture of herbivorous and omnivorous species. International Publication Number: WO2012/038892. 2012

[52] Pearce CM, Daggett TL, Robinson SMC. Effect of protein source ratio and protein concentration in prepared diets on gonad yield and quality of the green sea urchin, *Strongylocentrotus droebachiensis*. Aquaculture. 2002;**214**:307-332

[53] Robinson SMC, Castell JD, Kennedy EJ. Developing suitable colour in the gonads of cultured green sea urchins (*Strongylocentrotus droebachiensis*). Aquaculture. 2002;**206**:289-303

[54] Tomšić S, Conides AJ, Aničić I. Growth and gonad changes in stony sea urchin, *Paracentrotus lividus* (lamark, 1816) fed artificially formulated feed and benthic macrophyte diet. Naše more. 2015;**62**(2):85-90. DOI: 10.17818/NM/2015/2.7

[55] Basuyaux O, Blin JL. Use of maize as a food source for sea urchins in a recirculating rearing system. Aquaculture International. 1998;**6**:233-247

[56] Garmendia JM, Menchaca I, Belzunce MJ, Franco J, Revilla M. Induction to maturation of the sea urchin *Paracentrotus lividus* (Lamarck, 1816) under laboratory conditions. Environmental Technology. 2009;**30**(13, 1):1441-1446

[57] Codomier L, Bruneau Y, Combaut G, Teste J. Etude biologique et chimique d'Asparagopsis armata et de Falkenbergia rufolanosa (Rhodophycées, Bonnemaisoniales). CR Acad Sci Paris. 1977;**284D**:1163-1165

[58] Cabrita MT, Vale C, Rauter AP. Halogenated compounds from marine algae. Marine Drugs. 2010;**8**:2301-2317. DOI: 10.3390/md8082301

[59] McCarron E, Burnell G, Kerry J, Mouzakitis G. An experimental assessment on the effects of photoperiod treatments on the somatic and gonadal growth of the juvenile European purple sea urchin *Paracentrotus lividus*. Aquaculture Research. 2010;**41**:1072-1081

[60] Fabbrocini A,, Volpe MG, Di Stasio M, D'Adamo R, Maurizio D, Coccia E Paolucci M. Agar-based pellets as feed for sea urchins (*Paracentrotus lividus*): Rheological behaviour, digestive enzymes and gonad growth. Aquaculture Research. 2012;**43**:321-331

[61] Fernandez, C.; Boudouresque, CF. Evaluating artificial diets for small *Paracentrotus lividus* (Echinodermata: Echinoidea). In: Mooi R, Telford M, editors. Echinoderms: San Francisco. Rotterdam: Balkema; 1998. pp. 651-656

[62] Fernandez C, Boudouresque CF. Nutrition of the sea urchin *Paracentrotus lividus* (Echinodermata: Echinoidea) fed different artificial food. Marine Ecology Progress Series. 2000;**204**:131-141

[63] Spirlet C, Grosjean Ph, Jangoux M. Cultivation of *Paracentrotus lividus* (Echinodermata: Echinoidea) on extruded feeds: Digestive effciency, somatic and gonadal growth. Aquaculture Nutrition. 2001;**7**:91-99

[64] Schlosser SC, Lupatsch I, Lawrence JM, Lawrence AL, Shpigel M. Protein and energy digestibility and gonad development of the European sea urchin *Paracentrotus lividus* (Lamarck) fed algal and prepared diets during spring and fall. Aquaculture Research. 2005;**36**:972-982

[65] Fabbrocini A, D'Adamo R. Gametes and embryos of sea urchins (*Paracentrotus lividus*, Lmk., 1816) reared in confined conditions: Their use in toxicity bioassays. Chemistry and Ecology. 2011;**27**(sup 2):105-115. DOI: 10.1080/02757540.2011.625931

[66] Luis O, Delgado F, Gago J. Year-round captive spawning performance of the sea urchin *Paracentrotus lividus*: Relevance for the use of its larvae as live feed. Aquatic Living Resources. 2005;**18**:45-54

[67] Guidetti P, Terlizzi A, Boero F. Effects of the edible sea urchin, *Paracentrotus lividus*, fishery along the Apulian rocky coast (SE Italy, Mediterranean Sea). Fisheries Research. 2004;**66**:287-297

[68] Pais A, Chessa LA, Serra S, Ruiu A, Meloni G, Donno Y. The impact of commercial and recreational harvesting for *Paracentrotus lividus* on shallow rocky reef sea urchin communities in North-western Sardinia, Italy. Estuarine, Coastal and Shelf Science. 2007;**73**:589-597

[69] ICES. Report of the ICES Advisory Committee on the Marine Environment. 1997

[70] Pagano G, Cipollaro M, Corsale G, Esposito A, Ragucci E, Giordano GG, Trieff NM. Comparative toxicities of chlorinated biphenyls on sea urchin egg fertilization and embryogenesis. Marine Environmental Research. 1985;**17**:240-244

[71] Pagano G, Cipollaro M, Corsale G, Esposito A, Giordano GG, Ragucci E, Trieff NM. Comparative toxicities of benzene, chlorobenzene and dichlorobenzene to sea urchin embryos and sperm. Bulletin of Environmental Contamination and Toxicology. 1988;**40**: 481-488

[72] Pagano G, Anselmi B, Dinnel PA, Esposito A, Guida M, Iaccarino M, Melluso G, Pascale M, Trieff NM. Effects on sea urchin fertilization and embryogenesis of water and sediment from two rivers in Campania, Italy. Bulletin of Environmental Contamination and Toxicology. 1993;**25**:20-26

[73] Pagano G, Korkina LG, Iaccarino M, De Biase A, Deeva IB, Doronin YK, Guida M, Melluso G, Meriç S, Oral R, Trieff NM, Warnau M. Developmental, cytogenetic and biochemical effects of spiked or environmentally polluted sediments in sea urchin bioassays. In: Garrigues P, Walker CH, Barth H, editors. Biomarkers in Marine Ecosystems: A Practical Approach. Amsterdam: Elsevier; 2001. pp. 85-129. DOI: 10.1016/B978-044482913-9/50007-9

[74] Bressan M, Marin MG, Brunetti R. Effects of linear alkylbenzene sulphonate (LAS) on skeletal development of sea urchin embryos (*Paracentrotus lividus* LMK). Water Research. 1991;**25**:613-616. DOI: 10.1016/0043-1354(91)90134-C.

[75] Pieroni M, Falugi C. Effects of cholinergic drugs on cell interactions during fertilization and early development of the sea urchin *Paracentrotus lividus*. Bollettino Della Societa Italiana Di Biologia Sperimentale. 1992;**68**:16-22

[76] Graillet C, Pagano G, Girard JP. Stage-specific effects of teratogens on sea urchin embryogenesis. Teratogen Carcinogen Mutagen. 1993;**130**:1-14

[77] Volpi Ghirardini A, Birkemeyer T, Arizzi Novelli A, Pavoni B, Ghetti PF. An integrated approach to sediment quality assessment: The Venice lagoon as a case study. Aquatic Ecosystem Health & Management. 1999;**2**:435-447

[78] Volpi Ghirardini A, Arizzi Novelli A. A sperm cell toxicity test procedure for the Mediterranean species *Paracentrotus lividus* (Echinodermata: Echinoidea). Environmental Technology. 2001;**22**:439-445

[79] Volpi Ghirardini A, Arizzi Novelli A, Losso C, Ghetti PF. Sea urchin toxicity bioassays for sediment quality assessment in the Lagoon of Venice (Italy). Chemistry and Ecology. 2003;**19**(2-3):99-111

[80] Volpi Ghirardini A, Arizzi Novelli A, Tagliapietra D. Sediment toxicity assessment in the Lagoon of Venice (Italy) using *Paracentrotus lividus* (Echinodermata: Echinoidea) fertilization and embryo bioassays. Environment International. 2005;**31**:1065-1077

[81] Arizzi Novelli A, Argese E, Tagliapietra D, Bettiol C, Volpi Ghirardini A. Toxicity of tributyltin and triphenyltin to early life-stages of *Paracentrotus lividus* (Echinodermata: Echinoidea). Environmental Toxicology and Chemistry. 2002;**21**(4):859-864

[82] Arizzi Novelli A, Losso C, Ghetti PF, Volpi Ghirardini A. Toxicity of heavy metal using sperm cell and embryo toxicity with *Paracentrotus lividus* (Echinodermata: Echinoidea): Comparison with exposure concentration in the lagoon of Venice, Italy. Environmental Toxicology and Chemistry. 2003;**22**(6):1295-1301

[83] Sartori D, Scuderi A, Sansone G, Gaion A. Echinoculture: The rearing of *Paracentrotus lividus* in a recirculating aquaculture system. Experiments of artificial diets for the maintenance of sexual maturation. Aquaculture International. 2015;**23**:111-125

[84] Arizzi Novelli A,Picone M, Losso C, Volpi Ghirardini A. Ammonia as confounding factor in toxicity tests with the sea urchin *Paracentrotus lividus* (Lmk). Toxicological & Environmental Chemistry. 2003;**85**(4-6):183-191

[85] Russo R, Bonaventura R, Zito F, Schröder HC, Müller I, Müller WEG, Matranga V. Stress to cadmium monitored by metallothionein gene induction in *Paracentrotus lividus* embryos. Cell Stress & Chaperones. 2003;**8**:232-241

[86] Losso C, Arizzi Novelli A, Picone M, Volpi Ghirardini A, Ghetti PF, Rudello D, Ugo P. Sulfide as a confounding factor in toxicity tests with the sea urchin *Paracentrotus lividus*: Comparisons with chemical analysis data. Environmental Toxicology and Chemistry. 2004;**23**(2):396-401

[87] Angelini C, Aluigi MG, Sgro M, Trombino S, Thielecke H, Falugi C. Cell signalling during sea urchin development: A model for assessing toxicity of environmental contaminants. Progress in Molecular and Subcellular Biology. 2005;**39**:45-70

[88] Schröder HC, Di Bella G, Janipour N, Bonaventura R, Russo R,. Müller WEG, Matranga V. DNA damage and developmental defects after exposure to UV and heavy metals in sea urchin cells and embryos compared to other invertebrates. Progress in Molecular and Subcellular Biology. 2005;**39**:111-137

[89] Gaion A, Scuderi A, Pellegrini D, Sartori D. Arsenic exposure affects embryo development of sea urchin, *Paracentrotus lividus* (Lamarck, 1816). Archives of Environmental Contamination and Toxicology. 2013;**91**(5):565-570. DOI: 10.1007/s00128-013-1098-0

[90] Bunea A, Andjelkovic M, Socaciu C, Bobis O, Neacsu M, Verhe R, Van Camp J. Total and individual carotenoids and phenolic acids contentin fresh, refrigerated and processed spinach (*Spinacia oleracea* L.). Food Chemistry. 2008;**108**:649-656. DOI: 10.1016/j.foodchem.2007.11.056

[91] Lomnitski L, Carbonatto M, Ben-Shaul V, Peano S, Conz A, Corradin L, Maronpot RR, Grossman S, Nyska A. The prophylactic effects of natural water-soluble antioxidant from spinach and apocynin in a rat model of lipopolysaccharide-induced endotoxemia. Toxicologic Pathology. 2000;**28**:588-600

[92] Lomnitski L, Foley J, Grossman S, Ben-Shaul V, Maronpot R, Moomaw C, Carbonatto M, Nyska A. Effects of apocynin and natural antioxidants from spinach on iNOS and COX-2 induction in LPS-induced hepatic injury in rat. Journal of Pharmacology and Toxicology. 2000;**87**:18-25

[93] Nyska A, Lomnitski L, Spalding J, Dunson DB, Goldsworthy TL, Grossman S, Bergman M, Boorman G. Topical and oral administration of the natural water-soluble antioxidant from spinach reduces the multiplicity of papillomas in the Tg.AC mouse model. Toxicology Letters. 2001;**122**:33-44

[94] Rice-Evans CA, Miller NJ, Paganga G. Structure-antioxidant activity relationships of flavonoids and phenolic acids. Free Radical Biology & Medicine. 1996;**20**:933-956

[95] Dreosti IE. Antioxidant polyphenols in tea, cocoa, and wine. Nutrients. 2000;**16**:692-701

[96] Zhu L, Luo X, Jin Z. Effect of resveratrol on serum and liver lipid profile and antioxidant activity in hyperlipidemia rats. Asian Australasian Journal of Animal Sciences. 2008;**21**:890-895

[97] Verlaque M, Nédélec H. Biologie de *Paracentrotus lividus* (Lamarck) sur substrat rocheux en Corse (Méditerranée, France): Alimentation des adultes. Vie Milieu. 1983;**33**(3-4):191-201

[98] Verlaque M. Biologie des juveniles de l'oursin herbivore *Paracentrotus lividus*: Sélectivité du broutage et impact de l'espèce sur les communautés algales de substrat rocheux en Corse (Méditerranée, France). Botanica Marina. 1984;**27**(9):401-424

[99] Verlaque M. Relations entre *Paracentrotus lividus* (Lamarck) et le phytobenthos de Mediterranée occidentale. In: Boudouresque CF, editor. Colloque international sur *Paracentrotus lividus* et les oursin comestibles. Marseille: GIS Posidonie Publication; 1987. pp. 5-36

[100] Traer K. The consumption of *Posidonia oceanica* Delile by Echinoids at the Isle of Ischia. In: Jangoux M, editor. Echinoderms: Present and Past. Rotterdam: AA Balkema Press; 1980. pp. 241-244

[101] Dado RG. Nutritional benefits of specialty maize grain hybrids in dairy diets. Journal of Animal Science. 1999;**77**(2):197-207

[102] Griffiths M, Perrott P. Seasonal changes in the carotenoids of the sea urchin *Strongylocentrotus droebachiensis*. Comparative Biochemistry and Physiology. 1976;**55B**:435-441

[103] Matsuno T, Tsushima M. Carotenoids in sea urchins. In: Lawrence JM, editor. Edible Sea Urchins: Biology and Ecology. 2nd ed. Amsterdam: Elsevier Science BV; 2001. pp. 115-138

[104] Symonds R, Kelly M, Caris-Veyrat C, Young A. Carotenoids in the sea urchin *Paracentrotus lividus*: Occurrence of 90-cis-echinenone as the dominant carotenoid in gonad colour determination. Comparative Biochemistry and Physiology Part B: Biochemistry and Molecular Biology. 2007;**148**:432-444

[105] Krinsky NI. The biological properties of carotenoids. Pure and Applied Chemistry. 1994;**66**:1003-1010

[106] Nacci D, Jackim E, Walsh R. Comparative evaluation of three rapid marine toxicity tests: Sea urchin early embryo growth test, sea urchin sperm cell toxicity test and microtox. Environmental Toxicology and Chemistry. 1986;**5**:521-525

[107] Dinnel PA, Link JM, Stober QJ. Improved methodology for a sea urchin sperm cell bioassay for marine waters. Archives of Environmental Contamination and Toxicology. 1987;**16**:23-32

[108] Warnau M, Temara A, Jangoux M, Dubois P, Iaccarino M, De Biase A, Pagano G. Spermiotoxicity and embryotoxicity of heavy metals in the echinoid *Paracentrotus lividus*. Environmental Toxicology and Chemistry. 1996;**15**:1931-1936. DOI: 10.1002/etc.5620151111

[109] Fernández N, Beiras R. Combined toxicity of dissolved mercury with copper, lead and cadmium on embryogenesis and early larval growth of the *Paracentrotus lividus* sea-urchin. Ecotoxicology. 2001;**10**:263-271

[110] Lera S, Pellegrini D. Evaluation of the fertilization capability of *Paracentrotus lividus* sea urchin storage gametes by the exposure to different aqueous matrices. Environmental Monitoring and Assessment. 2006;**119**:1-13

[111] Sun C-K, Chu S-W, Chen S-Y, Tsai T-H, Liu T-M, Lin C-H, Tsai H-J. Higher harmonic generation microscopy for developmental biology. Journal of Structural Biology. 2004;**147**:19-30

[112] Chen S-Y, Hsieh C-S, Chu S-W, Lin C-H, Ko C-Y, Chen Y-C, Tsai H-J,. Hu C-H, Sun C-K. Noninvasive harmonics optical microscopy for long-term observation of embryonic nervous system development in vivo. Journal of Biomedical Optics. 2006;**11**:054022-054022-8. DOI: 10.1117/1.2363369

[113] Buttino I, Hwang J-S, Sun C-K, Hsieh C-T, Liu T-M, Pellegrini D, Ianora A, Sartori, D, Romano G, Cheng S-H, Miralto A. Apoptosis to predict copepod mortality: State of the art and future perspectives. Hydrobiologia. 2011;**666**:257-264

[114] Rehberg M, Krombach F, Pohl U, Dietzel S. Label-Free 3D visualization of cellular and tissue structures in intact muscle with second and third harmonic generation microscopy. PLoS ONE. 2011;**6**(11):e28237. DOI: 10.1371/journal.pone.0028237

[115] Hsieh C-S, Ko C-Y, Chen S-Y, Liu T-M, Wu J-S, Hu C-H, Sun C-K. In vivo long-term continuous observation of gene expression in zebra fish embryo nerve system by using harmonic generation microscopy and morphant technology. Journal of Biomedical Optics. 2008;**13**:064041-1-064041-7

[116] Sartori D, Pellegrini D, Macchia S, Gaion A. Can echinoculture be a feasible and effective activity? Analysis of fast reliable breeding conditions to promote gonadal growth and sexual maturation in *Paracentrotus lividus*. Aquaculture. 2016;**451**:39-46

[117] Valori-alimenti.com© 2007-2017 [Internet]. Available from: http://www.valori-alimenti.com/ [Accessed: March 6, 2017]

[118] Lester GE, Makus DJ, Hodges DM. Relationship between fresh-packaged spinach leaves exposed to continuous light or dark and bioactive contents: Effects of cultivar, leaf size, and storage duration. Journal of Agricultural and Food Chemistry. 2010;**58**(5): 2980-2987

[119] Golden T, Hinerfeld DA, Melov S. Oxidative stress and aging: Beyond correlation. Aging Cell. 2002;**1**:117-123. DOI: 10.1046/j.1474-9728.2002.00015.x

[120] Lizarazo K, Fernández-Marın B, Becerril JM, García-Plazaola JI. Ageing and irradiance enhance vitamin E content in green edible tissues from crop plants. Journal of the Science of Food and Agriculture. 2010;**90**:1994-1999. DOI: 10.1002/jsfa.4043

[121] Buttino I, Hwang JS, Romano G, Sun CK, Liu TM, Pellegrini D, Gaion A, Sartori D. Detection of malformations in sea urchin plutei exposed to mercuric chloride using different fluorescent techniques. Ecotoxicology and Environmental Safety. 2015;**123**:72-80. DOI: 10.1016/j.ecoenv.2015.07.027i

[122] Chiarelli R, Agnello M, Roccheri MC. Sea urchin embryos as a model system for studying autophagy induced by cadmium stress. Autophagy. 2011;**7**(9):1028-1034. DOI: 10.4161/auto.7.9.16450

[123] Agnello M, Bosco L, Chiarelli R, Martino C, Roccheri MC. In: Ntuli TM, editor. Cell Death—Autophagy, Apoptosis and Necrosis. The Role of Autophagy and Apoptosis During Embryo Development. Rijeka: InTech; 2015. pp. 83-112. DOI: 10.5772/61765

[124] Roccheri MC, Tipa C, Bonaventura R, Matranga V. Physiological and induced apoptosis in sea urchin larvae undergoing metamorphosis. International Journal of Developmental Biology. 2002;**46**(6):801-806

[125] Raposo A, Ferreira SM, Ramos R, Anjos CM, Baptista T, Tecelao C, Goncalves SC, Pombo A. Effect of three diets in the gonadal growth and maturation of *Paracentrotus lividus*. Frontiers in Marine Science Conference Abstract: IMMR | International Meeting on Marine Research; 2016. DOI: 10.3389/conf.FMARS.2016.04.00030

[126] Sanja T, Conides A, Anicic I. Growth and gonad changes in stony sea urchin, *Paracentrotus lividus* (Lamark, 1816) fed artificially formulated feed and benthic macrophyte diet. Nase More. 2015;**62**(2):85-90. DOI: 10.17818/NM/2015/2.7

[127] Colak SO, Canak O, Balta G. Sea urchin (*Paracentrotus lividus*) gonadal histology. In: Proceedings of the International Symposium on Fisheries and Aquatic Sciences FABA; 3-5 November 2016; Antalya, Turkey

[128] Sartori D, Pellegrini D, Gaion A. Analysis of variability in embryological response of two sea urchin species to spatial and temporal features-can these factors influence responses in standardized ecotoxicological assays? Expert Opinion on Environmental Biology. 2016;**S1**. DOI: 10.4172/2325-9655.S1-002.

[129] Colin H, Officer RA, Chamberlain J. Evaluation of the efficacy of algal conditioned substrates for inducing settlement of *Paracentrotus lividus* larvae. Aquaculture Research. 2015;1-6. DOI: 10.1111/are.12959

[130] Brundu G, Monleón LV, Vallainc D, Carboni S. Effects of larval diet and metamorphosis cue on survival and growth of sea urchin post-larvae (*Paracentrotus lividus*; Lamarck, 1816). Aquaculture. 2016;**465**:265-271

Section 4

Sea Urchin Suitability for Biomedical Studies

Chapter 6

The Sea Urchin Embryo: A Model for Studying Molecular Mechanisms Involved in Human Diseases and for Testing Bioactive Compounds

Maria Di Bernardo and Marta Di Carlo

Additional information is available at the end of the chapter

http://dx.doi.org/10.5772/intechopen.70301

Abstract

Most of the current knowledge concerning fundamental genetic mechanisms, evolutionary processes and development, cellular physiology, and pathogenesis comes from studies of different animal model systems. Whereas mice, rats, and other small mammals are generally thought of as the typical model systems used by researchers in biomedical studies, aquatic models including both freshwater and marine organisms have long proved to be essential for the study of basic biological processes. For over a century, biologists have used the sea urchin embryo as a prototype for the investigation of developmental mechanisms that contribute to building the embryo body plan. Here we highlight the contribution of the sea urchin embryo as a simple model for studying aging and age-associated neurodegenerative diseases, as well as the pathways and mechanisms involved in cell survival and death. Moreover, we point out the role of this embryonic system as a potent and affordable tool for learning about developmental effects and toxicity responses to environmental contaminants and chemical compounds.

Keywords: embryonic model system, developmental effects, neurodegeneration, senescence, chemical compound assays

1. Introduction

Sea urchins are well-studied marine organisms belonging to the Echinoderm phylum. Since the mid-nineteenth century, the amazing transparent sea urchin embryo has been one of the favored animal models for descriptive experimental work on embryo development.

Afterward, it has been used for studying intercellular communication and cell adhesion [1–3], cell cycle control mechanisms [4], calcium signaling [5], fertilization [6], cell differentiation [7], and cell survival and death [8].

Added benefits of this system are low maintenance costs, small size, high fecundity, and the transparency of embryos, features that allow the direct observation of cell division and movement inside the living embryos and larvae. Moreover, as invertebrate species, sea urchins are not subject to restricted animal welfare concerns. Notably, this trait meets the strategy of the European Partnership for Alternative Approaches to Animal Testing for the development of alternative approaches to using animals in biological assays.

Studies on different sea urchin species have, indeed, identified maternal molecules which are spatially restricted and involved in the determination of cell fate [9–12], molecular mechanisms that respond to cellular stress such as heat shock [13–15], apoptosis [16], and autophagy [17]. More recently, researchers have used sea urchin embryos as a model for elucidating the role of cellular and molecular mechanisms involved in human health and disease. The potential of using the sea urchin as a model for disease research relies on the fact that general cellular properties are common to many organisms. The complete sequencing of the sea urchin genome has also revealed that sea urchins are more closely related to humans than other invertebrates [18, 19], including the model organisms *D. melanogaster* and *Caenorhabditis elegans* which are commonly used as disease models. Research on the sea urchin animal model now extends over a wide range of areas, such as immunology, microbiology, pathology, toxicology, and microbiology.

Recently the sea urchin embryo has played an important role as a model in the study of neurodegenerative disorders that can cause dementia and memory loss. This is possible because of the presence of a larval nervous system that arises during gastrulation within the ectoderm of the embryo [20]. Later, a set of neurons and neurites begins to be present in the structure called the ciliary band. Several clusters of neurons with associated neuropil are organized in ganglia, the largest of which is the apical organ of the larva, composed of some bilaterally positioned sensory cells containing serotonin. Thus, dysfunction in the cellular processes of these cells can be compared to what occurs in human neurons during pathological dysfunctions. In neurodegenerative disorders, such as Alzheimer's disease, neuronal loss occurs due to a buildup of a particular protein called Amyloid beta (Abeta) that, in this pathological condition, is misfolded and prone to aggregating, forming fibrils and plaques [21]. Abeta is a peptide derived from the proteolysis of a membrane-spanning precursor protein called amyloid precursor protein (APP); to strengthen the view that sea urchins can be utilized in these studies, an antigen related to human APP was identified [22, 23].

Another interesting opportunity presented by sea urchins is the possibility of gaining new insight into aging mechanisms. Aging is a multifactorial process, and many theories, such as telomere loss, oxidative stress, and free radical theories, have been proposed to explain this phenomenon at the molecular, cellular, systemic, and evolutionary levels [24]. Some reports indicate that different life spans have been found for different sea urchin species ranging between few years and 100 years. The sea urchin *Strongylocentrotus franciscanus*, indeed, shows a negligible senescence with decreased mortality and retained reproductive capacity for a long time [25]. In contrast, *Lytechinus variegatus* has an estimated life span of only 4 years [26], while *Strongylocentrotus purpuratus* has an estimated maximum life span of more than 50 years [27]. Thus, to study sea urchin species with different life spans could be relevant for understanding mechanisms involved in aging.

Animal models also give an invaluable contribution to our knowledge of all biological aspects and processes, including the discovery and development of new chemical compounds by studying their properties and functions. The researcher's first choice is often to select models that are highly similar in biology and physiology to humans, although a combination of experimental data from different models can be more informative than those from a single organism, even evolutionarily closer. In vivo high-medium-throughput assays using small aquatic model organisms provide significant advantages for bioassay development, generating molecular and physiologically relevant responses. Among them, the sea urchin embryo is considered a valuable model to test and assess the safety of biomolecules, even those developed for therapeutic purposes [28–31] but also to discover basic molecular mechanisms and gene regulatory networks with which they may interfere.

In this chapter, we describe the role of the sea urchin embryo in studies of aging and age-related neurodegenerative disorders. Furthermore, we provide an update and underline the relevance of this model for assessing the developmental responses and/or toxicity effects of small molecules, many of which are used for medical purposes.

2. The sea urchin as a tool for studying neurodegenerative disorders

Neurodegenerative diseases are defined as hereditary and sporadic conditions, which are characterized by the progressive dysfunction of central or peripheral structures of the nervous system. These disorders are often debilitating disorders affecting memory, learning, skilled movements, and feelings. They include common diseases such as Alzheimer's disease (AD) and Parkinson's disease (PD) and rarer disorders such as multiple sclerosis, amyotrophic lateral sclerosis (ALS), Huntington's disease (HD), spinocerebellar ataxia, and prion diseases. Even though their clinical manifestations and symptoms are completely different, neurodegenerative disorders share some common features, such as their appearance with aging, neuronal loss and synaptic anomalies, and mainly the presence of cerebral deposits of misfolded protein aggregates [32]. Although the presence of extracellular deposits is a common feature of these diseases, the protein component and distribution of the aggregates are specific to each pathology. Thus, for example, β-amyloid is the main component of the deposits in AD, α-synuclein in PD, and polyglutamine-rich in HD, and the accumulation of the protease-resistant aggregates of the prion protein (PrP) is present in the diverse forms of transmissible spongiform encephalopathy. The presence of abnormal aggregates in damaged regions of the brain is due to protein misfolding, which is one of the main causes of neurodegenerative diseases. Generally, the specific native protein is mainly composed of an unordered α-helical structure, but in the pathological condition, it is misfolded in a β-sheet structure that aggregates to form ordered protofibrils, fibrils, and plaques. Today, it is widely accepted that small oligomers, rather than compacted aggregates deposited in the brain, are the real causes of neurodegeneration [33, 34]. However, regardless of the different sizes of aggregates, the induced toxicity leads to different cellular dysfunctions that culminate in the neuronal apoptotic process, though the mechanism by which protein misfolding and aggregation trigger neurodegeneration is still unclear. Since molecular mechanisms and basic pathways are conserved during evolution, the use of a simple model system can help us to understand the derived dysfunction and the pathways affected by a specific insult. Among the neurodegenerative diseases, AD is a pathology that is rapidly increasing with longer human life expectancy. AD is, indeed, the most common cause of dementia in elderly people and typically begins with subtle and weakly recognized failures of memory and slowly becomes more severe. It leads to the progressive loss of mental, behavioral, and functional abilities. The pathological hallmarks in the AD brain present two different abnormal structures, extracellular amyloid plaques, composed of amyloid-β (Aβ) peptides, and intraneuronal neurofibrillary tangles, composed of hyperphosphorylated tau [21]. Amyloid-β peptides (Aβ) are proteolytic fragments of 40–42 amino acids, derived from the transmembrane amyloid precursor protein (APP), whereas tau is a brain-specific, axon-enriched microtubule-associated protein. Depending on cellular conditions, Aβ is misfolded, and the establishment of ordered structures rich in β-sheet, prone to self-assembling, produces aggregates typical of amyloid assemblies. For a long time, large aggregates of Aβ, found in amyloid plaques, have been considered the major cause of neuron damage and degeneration in AD. Recent studies have, instead, demonstrated that Aβ soluble oligomers, also known as Abeta-derived diffusible ligands (ADDL), are the more toxic form [35]. The combination of biochemistry, molecular and cell biology, and systems biology has been utilized for understanding some of the main molecular mechanisms leading to AD onset, but other efforts and knowledge are required. AD is probably caused by complex interactions among multiple genetic, epigenetic, and environmental factors; thus, interdisciplinary approaches leading to the identification of useful biomarkers are necessary for the development of suitable drugs for prevention and treatment.

In this scenario, how can a simple model system help AD neurodegeneration research? What questions can we ask and what answers can we have from a simple model system? For this aim, the first issue to be addressed is to obtain evidence that Aβ induces toxicity on the chosen model system. As mentioned earlier, the sea urchin is a suitable model system to test effects induced by chemical or natural toxic agents on biochemical pathways and embryo morphology. The sea urchin has a primordial nervous system, and the same neurotransmitters used in embryonic development are conserved between sea urchins and the mammalian brain. The sea urchin embryo, indeed, synthesizes, stores, and releases acetylcholine (Ach) and other neurotransmitters and possesses analogous receptors and downstream signaling cascades, all of which appear over a well-defined developmental period and act as morphogens in controlling cell differentiation and embryo and larva assembly [36]. Furthermore, under specific stimuli, the sea urchin embryo triggers apoptosis [16], the death mechanism by which neurodegeneration occurs (**Figure 1**). Studies from different laboratories have utilized two sea urchin species, *Paracentrotus lividus* and *Sphaerechinus granularis*, to test Aβ toxicity.

By using the sea urchin as model system, one of the first questions to be addressed was to establish if different structural Aβ species have the same toxicity. To this aim, it was produced a recombinant $A\beta_{42}$ peptide by using the pQ30 expression vector [37]. Different sizes

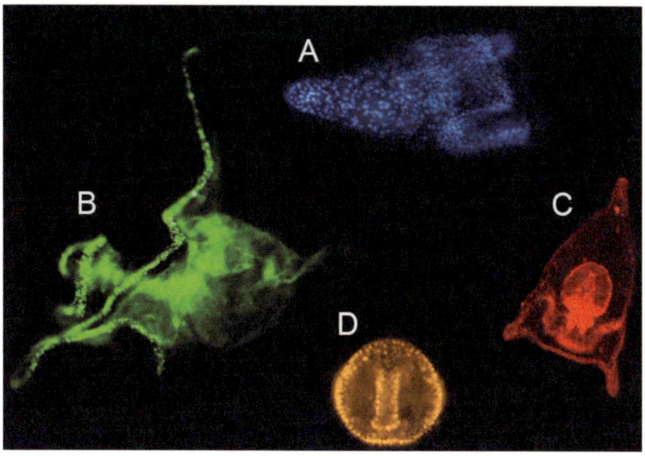

Figure 1. Sea urchin embryos have neurons and neurites, and respond to classical mechanisms involved in neurodegeneration processes, such as apoptosis or oxidative stress. *P. lividus* gastrula and pluteus stage embryos are depicted. (A, D), nuclei of pluteus and gastrula stage embryos are stained with Hoechst 33258; (B), late-stage larva stained with anti U-295 antibody (kind gift of Dr. David R. McClay) that specifically recognizes neurites at the ciliary band; (C) reactive oxygen species detected by a fluorescent dye in a pluteus stage embryo.

of aggregates, either small oligomers or larger aggregates, were obtained upon dissolving the recombinant Aβ_{42} under different pH conditions (7.2 or 3, respectively); different ionic strengths, indeed, change the kinetic of oligomer formation. By using a light scattering instrument, it was established that at pH 7.2, small Aβ_{42} species were present in solution; in contrast, at pH 3, larger aggregates were detected. To define a structure-toxicity relationship, Aβ_{42} was dissolved in the two solutions and administered, in different concentrations, to a two-cell-stage embryo culture of *Paracentrotus lividus*. Under these conditions, embryos were allowed to develop until controls had reached the pluteus stage, and different percentages of surviving or degenerating embryos, with respect to the controls, were observed. After only 4 h of embryo development, morphological defects in the cell membrane were observed. After 1 day of development, instead, retardation of embryo development as well as cellular disorders visible inside the blastocoele was detectable. After 48 h of development, cellular degeneration in two different pathological phenotypes, the occluded blastulae and the occluded prism, was present [37]. In general, small oligomers were more toxic than larger aggregates, in agreement with the discovery that small ADDLs are highly toxic [35], indicating that the Aβ_{42} state of assembly appeared to influence its biological activity. Based on the obtained results, it was supposed that more diffusible small oligomers could be more easily internalized than larger aggregates in the developing embryo, producing cellular and molecular dysfunction. Thus, the sea urchin was a suitable in vivo organism for Aβ_{42} toxicity studies and could be used as an indicative tool for the pharmacological evaluation of a therapeutic agent.

Regarding Aβ_{42} toxicity, another group identified both the critical periods in which different types of anomalies are induced by Aβ_{42}, and the protective role played by acetylcholine and other neurotransmitters such as serotonin and cannabinoids [38]. The sensitivity to Aβ_{42} is greatest when the peptide is introduced at the mid-blastula stage, in a particular period called "mid-blastula transition," when new genes become expressed by the embryonic genome. Thus, the neurotoxic effects of Aβ are associated with changes in gene transcription. Further, the morphological changes correspond to a level of toxicity indicating that sea urchins provide a system that allows for the rapid screening of potential therapeutic interventions [38]. Furthermore, the morphological anomalies were inhibited by the addition of lipid-permeable analogs of acetylcholine (arachidonoyl dimethylaminoethanol), serotonin (arachidonoyl serotonin), and cannabinoids (arachidonoyl vanillylamine), indicating that they can prevent the neurotoxicity associated with Aβ and be used for therapies that enhance cholinergic function, as well as for AD.

By using *S. granularis*, the effect of a specific 15-amino acid sequence (NWCKRGRKQCKTHPH) placed in positions 96–100 within the extracellular domain of the APP protein (APP_{96-110}) was evaluated and compared to that of $A\beta_{42}$ [39]. This fragment corresponds to a proteoglycan-binding domain that specifically controls neurite outgrowth and other aspects of neurodevelopment [40]. Sea urchin embryos at the 2–4 cell stage, blastula stage, or at late gastrula to early pluteus stages were treated with the two peptides, with or without neurotransmitters or neurotransmitter-related agents, and embryonic malformations were observed. With respect to $A\beta_{42}$, APP_{96-110} had a weaker effect in disrupting development, requiring higher concentrations to produce the same malformations caused by $A\beta_{42}$. APP_{96-110} was dangerous only within a defined window of vulnerability corresponding about to the mid-blastula stage, whereas $A\beta_{42}$ had adverse effects from early cleavage to the pluteus stage [39]. For both peptides, developmental anomalies were prevented or reduced by the addition of neurotransmitters. This finding indicated a role for APP in development and identified specific interactions with neurotransmitter systems that act as morphogens in developing sea urchins as well as in the mammalian brain.

3. The APP gene is conserved in the sea urchin genome

As mentioned above, Aβ is a peptide derived from a multiple proteolytic cleavage at the C-terminal position of a large transmembrane protein called APP [21]. To validate the sea urchin embryo as a model system in the study of Aβ-induced toxicity, it was necessary to demonstrate the presence of an antigen related to human APP. APP is a protein conserved during evolution, and APP-related genes have also been found in less evolved organisms, such as *Drosophila melanogaster* and the worm *Caenorhabditis elegans*. This suggested that common mechanisms and responses to pathological stimuli could be conserved [41–43]. Thus, the possible presence of an antigen related to APP in sea urchins was investigated. For this aim, the presence of APP was examined in proteins extracted from gastrula stage embryos and pluteus larvae at 48, 120, and 288 h of development [22]. The choice of these developmental stages was based on the moment at which a primordial nervous system appears. The sea urchin larval nervous system consists of an array of neurons that control swimming and feeding, and a more defined organization begins to be present at the late gastrula stage [20].

The presence of an antigen related to APP, called *Paracentrotus lividus* APP (*Pl*APP), was detected, and in addition, a fragment of lower molecular weight was identified at late pluteus stage, suggesting that after the gastrula stage, a portion of the *Pl*APP was proteolytically cleaved, producing a peptide as occurs in higher organisms. Immunofluorescence colocalization experiments with serotonin, a neuronal antigen, confirmed that *Pl*APP was present in sea urchin primordial neurons [22]. Furthermore, clear proof of the presence of a gene encoding for *Pl*APP was obtained by cloning and sequencing a full-length cDNA.

Subsequently, the possibility that *P. lividus* embryos could trigger different apoptotic pathways was investigated after stimulation with Aβ aggregates of different sizes. Aβ-induced apoptosis generally occurs through caspase-dependent pathways, even if caspase-independent pathways have also been described [44, 45]. In sea urchin embryos, apoptosis is never found during cleavage stages. It begins to manifest between the early blastula and late gastrula stages [16, 46]; thus, we performed apoptosis analysis only in embryos after the gastrula stage. In light of this evidence, *Paracentrotus lividus* embryos at the two-cell stage were incubated with small oligomers or fibrillar aggregates and cultured until the controls arrived at gastrula and pluteus stages [22]. By microscopic inspection, a higher percentage of malformed or dead embryos was observed upon treatment with Aβ oligomer forms with respect to the Aβ aggregate forms. By TUNEL assay, apoptosis on surviving embryos treated with the different aggregation forms was detected. In addition, it was found that aggregates employed an exclusively extrinsic apoptotic pathway via caspase 9 activation, whereas oligomers activated both extrinsic and intrinsic apoptotic pathways via caspase 9 and caspase 8, respectively. These findings suggested that a part of the smaller soluble oligomers was able to penetrate the cells and produce mitochondrial damage that activated caspase 9. In contrast, other oligomers could attach to a neuron surface-specific binding site and, perhaps, seed the aggregation process, mimicking what occurs in human pathology. In summary, these studies of the sea urchin embryo give information about the possibility that small oligomers may cross the membrane and penetrate the intracellular environment. This appears to be in agreement with biophysical and modeling studies that have demonstrated the ability of Aβ to interact with a lipid bilayer due to its obliquity and hydrophobicity [47].

4. Species-specific sea urchin longevity and senescence

Aging, also called senescence, is the process of becoming older. The term is particularly used to refer to humans, but all living organisms generally share this process. It is a well-defined biological route characterized by a progressive functional decline and increasing death over time. Because aging and life span characteristics diverge extremely between similar species, it has been believed that intrinsic organism life span is genetically determined and developed through an evolutionary process similar to the one that determines other species-specific characteristics. Furthermore, many theories have been proposed to explain this phenomenon at different levels such as loss of telomeres or enhanced free radical generation [24]. Telomeres are chromosome regions made up of repetitive sequences which function to protect chromosome deterioration. During aging, telomeres become shorter, and their attrition can be counteracted by telomerase activity. Thus, telomere attrition contributes to aging. The free radical, or oxidative stress, theory of aging, instead, proposes that the accumulation of oxidative cellular damage is a major contributor to the aging process and a key determinant of species longevity. The latter, referred to especially long-lived members of a population, is the challenger of aging. Studies on human longevity suggest that some of it is attributable to genetic factors and the rest is influenced by epigenetic and environmental factors [48]. Lifestyle and nutrition influence longevity at all stages of development and levels of human diversity [49, 50].

However, there are a number of different animals that show slower senescence, with decreased mortality and no reduction in reproductive capacity, no increase physiological dysfunction, and increased disease resistance with age [51, 52]. The study of these animals may furnish new insights about effective defenses against the degenerative process of aging. The sea urchin offers an ideal model to investigate mechanisms of longevity and reduced senescence. Different species of sea urchins, indeed, have very different natural life spans, ranging from 4 to more than 100 years, thus providing a unique model to investigate the molecular, cellular, and physiological mechanisms underlying both life span determination and negligible senescence. As told before, the species *Strongylocentrotus franciscanus* lives in excess of 100 years [25, 53]; *L. variegatus* has a life span of only 4 years [26], while *S. purpuratus* has an estimated life span of more than 50 years [27]. Identification of molecules involved in specific pathways that could be activated or inhibited in species with different longevity may provide insight into mechanisms involved in senescence. Thus, sea urchins represent an interesting alternative model for aging research [52]. To demonstrate that the red sea urchin, *S. franciscanus*, is a long-lived organism, some studies were carried out in two different localities of Washington State, USA [53]. By using a chemical marker, tetracycline-HCl, that binds to calcium ions and becomes incorporated into the skeleton during calcification [54], growth rates were determined. Further, since aging can be defined as the time-related deterioration of the physiological functions necessary for survival and fertility, these parameters were analyzed. Some individuals were tagged, and diameters (diameter test) of all sea urchins and their gonads, as survival and fertility parameters, were measured at the time of tagging as well as when collected after a year or more in the field, to establish the size structure of the populations. Gonad size, indeed, changes during the seasons, reaching a maximum in spring and minimum in summer, and these different sizes are correlated to reproductive capacity.

No decrease in gonad size with respect to the increasing diameter of individuals was found. By applying a mathematical model, change in survival and increasing reproductive esteems were done, and it was concluded that *S. franciscanus* shows no evidence of senescence. This was in contrast with the disposable soma theory [55], in which generally organisms have a limited amount of metabolic resources that have to be used to maintain the reproductive and nonreproductive activities of the organism (soma), including the repair of cellular damage. Senescence occurs when metabolic resources are exhausted and survival mechanisms, which operate throughout life, are altered. The data obtained on *S. franciscanus*, the long-lived species, indicate that it is subject to negligible or slow senescence [51] or none at all. As related above, life span differences among different species of the same organism have been found. This means that the biological changes leading to senescence occur at different ages, and comparison among long-lived (*S. fraciscanus*), intermediate-lived (*S. purpuratus*), and short-lived (*L. variegatus*) species could give new insights regarding the molecules, mechanisms, and key cellular pathways involved in life span determination and aging.

With the aim of finding the biomarkers involved in the longevity/senescence process, a proteomic study has been carried out using coelomic fluid of the three sea urchin species [56], whose age was determined by diameter test. Similarly to the blood, the coelomic fluid of sea urchins contains a miscellany of cells and macromolecules that provide essential functions such as nutrient transport as well as immune and clotting activities. Further, it contains proteins that are actively secreted as well as proteins released through cell lysis and cellular turnover. Proteomic analysis of the three sea urchin species revealed that the ectodomain of low-density lipoprotein receptor-related protein 4 (LRP4) was among the proteins that mainly increased with age. It has been proposed that since LRP4 is considered to be involved in Wnt signaling, the role of Wnt in negligible senescence should be better investigated [56].

5. Senescence and oxidative stress

According to the oxidative stress theory, both of the processes of aging and longevity are regulated by the accumulation of reactive oxygen species (ROS). ROS are partially reduced derivatives of oxygen that are highly reactive with major cell components such as proteins, lipids, and DNA, causing their damage. Many physiological cellular processes generate ROS, but others are produced by exposure to various external stimuli, such as ultraviolet light, ionizing radiation, and environmental toxins. Oxidative stress results from an imbalance between the production of ROS and the cell's intrinsic ability to inhibit damage through the production of antioxidant molecules or mechanisms that repair or eliminate damaged molecules. Thus, it is an imbalance between oxidants and antioxidants in favor of oxidants, which probably leads to damage. The existence of sea urchin species with different natural life spans, including some species with extraordinary longevity and negligible senescence, represents a model to study the accumulation of cellular oxidative stress with age. Oxidative cellular damage, antioxidant capacity, and proteasome enzyme activities were measured in different tissues of the three sea urchin species with different life spans, *L. variegatus, S. purpuratus,* and *S. franciscanus* [57]. No aged-related change in the marker of oxidative damage was observed in tissues of sea urchins with different life spans. Furthermore, levels of 8-hydroxy-2-deoxyguanosine, a marker of oxidative DNA damage, measured in cell-free coelomic fluid showed no general increase with age. Thus, the results suggested that negligible senescence is accompanied by a lack of accumulation of cellular oxidative damage with age; thus, the maintenance of antioxidant capacity and proteasome enzyme activities may be an important mechanism to mitigate time-linked damage.

Oxidative stress and mitochondrial dysfunction are the basis of aging and, consequently, to neurodegenerative diseases including AD. Thus, the use of natural antioxidants suppressing or reducing oxidative stress could be a neuroprotective strategy for blocking cell death. Ferulic acid (FA) (4-hydroxy-3-methoxycinnamic acid) is an antioxidant naturally present in plant cell walls. It has a phenolic nucleus and a long side chain, so it readily forms a resonance-stabilized phenoxy radical with high antioxidant [58] and anti-inflammatory [59] activities. It has been suggested that FA can act as a free radical scavenger [60]. Ferulic acid was used in vitro to block damage induced by beta amyloid, in the presence or not of drug delivery systems that could improve its transport and release in an in vivo system [61, 62]. *P. lividus* embryos were used as models to test the ability of FA to reduce cytotoxicity induced by Abeta. By morphological and fluorimetric test, it was possible to demonstrate that FA reduced the number of perturbated embryos induced by $A\beta_{42}$ [63]. Under this stimulus, sea urchins induced ROS generation and mitochondrial dysfunction, leading to embryo degeneration, and this process was counteracted by FA addition. Furthermore, after $A\beta_{42}$ treatment, a modulation of the molecules involved in the apoptosis process and activated by oxidative stress, such as the transcription factor foxo3a, was observed. In agreement with its antioxidant role, FA was able to inhibit the degenerative process through the downregulation of foxo3a. As happens in higher organisms, the sea urchin embryo exploits molecules, pathways, and mechanisms involved in both survival and death processes.

6. Sea urchin cells and embryos respond to external agents: humans exploit their skills!

Echinoderms, together with the Hemichordates, are sister groups of the Chordate phylum which includes humans and other vertebrates. As revealed by the analysis of the genome of the sea urchin *S. purpuratus*, the first nonchordate marine deuterostome to be completely sequenced [18, 64–66], a significant amount of the sea urchin gene repertoire is genetic material exclusive to the deuterostome superclade. Notably, a large part of the human gene catalogue, including orthologs of many human disease-associated genes, is expressed in sea urchins. Genome drafts, together with transcriptomes and other expression data from sea urchins and other Echinoderm species, are available at the SpBase and Octopus websites (http://www.echinobase.orghttp://www.echinobase.org/Echinobase/ and http://octopus.obs-vlfr.fr/, respectively).

A large number of studies have highlighted the power of the sea urchin embryonic system. The discovery of conserved gene regulatory networks (GRNs) driving development and morphogenesis has been a milestone in biology. Interactions between sequence-specific DNA binding molecules and DNA regulatory regions, together with signaling interactions and cofactors, control transcription and gene spatial temporal and expression [67–71]. Development, cell-type specification, and differentiation depend on these hardwired processes. A canonical and widespread developmental event, the embryonic epithelial mesenchymal transition (EMT), has been intensely studied in sea urchins. Primary mesenchyme cell (PMC) ingression during embryogenesis is evolutionarily conserved and is an excellent model of EMT in vivo [72, 73]. In humans, reactivated EMT drives organ fibrosis and tumor progression [74–76]. The approach to the identification of the GRN and subcircuits controlling EMT in sea urchin embryos constitutes a beautiful example of the strength of this system, as an excellent model system for the analysis of the transcription factors controlling EMT [77]. In addition, more traditional approaches have helped to discover networks and signaling pathways and even highlight the contribution of single molecules participating in embryo patterning along the embryonic axes [78–86].

The US National Institutes of Health has designated the sea urchin embryo as a model system helping to explain processes related to human health and disease [18]. Medical research and studies on environmental contaminants rely on the development of simple and affordable bioassays in cell systems and model organisms. These experimental activities are essential to elucidate mechanisms through which chemical compounds may exert their effects. Sea urchin embryo assays show high sensitivity and experimental reproducibility, providing rapid evidence of gamete abnormalities, developmental defects, and molecular changes in gene transcription and signaling pathways involved in constructing the embryo. The activity of several chemical compounds has been evaluated using sea urchin embryos along their early developmental stages, allowing the use of hundreds or even thousands of embryos in single screens. The high-throughput properties of this system benefit from the large amount of molecular and evolutionary data in the literature joined to the availability of powerful molecular tools that researchers of the sea urchin scientific community exchange each other [87].

Sea urchins express gene families that participate in response to environmental stressors. A genome-wide survey of the chemical defense gene network in *S. purpuratus* revealed around 400 genes, whose products contribute to the response to environmental stressors and cellular homeostasis. They include genes encoding for cytochrome oxidases, conjugating enzymes, ATP-dependent efflux transporters, oxidative detoxification proteins, and transcription factors involved in their regulation, many of which are expressed during embryogenesis. A great part of this repertoire is extraordinarily conserved during evolution [88, 89].

Members of the ATP-binding cassette (ABC) superfamily transport-specific molecules and substrates across membranes and mutations in these genes contribute to several human genetic disorders [90]. In sea urchin embryos, efflux transporter genes act as a first line of defense against toxic xenobiotic compounds; their expression is upregulated after fertilization [91], and this process is accomplished with relatively small energy costs [92, 93]. Cytochrome P450 family (CYPs) enzymes catalyze the oxidation of many xenobiotics, environmental chemicals, and drugs and protect the embryo from toxicants. Cytochrome P450-dependent oxidase activities have been used as potential markers for the assessment of environmental quality in marine areas, using different Mediterranean species including the sea urchin *P. lividus* [94].

In the sea urchin embryo, inducible forms of the heat shock proteins HSP70 (HSP70/HSP72) mainly mediate the cellular stress response. In sea urchin embryos, HSPs were first identified after heat treatment [14], although their expression has also been induced by heavy metals [95], UVB radiation, and other environmental pressures [96, 97]. It has been shown that proteins involved in stress responses cross talk with the immune signaling pathways [98]. In contrast to mammals that use both adaptive and innate immune responses, sea urchins only developed innate immune functions, as testified by the presence of an immune system with a rich repertoire of recognition receptors, regulators, and effectors [99]. This complex repertoire provides a nonspecific response to infection and/or injury [100, 101]. Immune receptor genes include molecules participating in the recognition of pathogens. Toll-like receptors (TLRs) and the echinoid-specific 185/333 gene family are expressed in different classes of adult immune cells, known as coelomocytes, while NACHT domain-LRR (NLR) cytoplasmic receptors have been found in the gut epithelium. Another expanded gene family present in the sea urchin genome encodes for the scavenger receptor cysteine-rich (SRCR) proteins [89, 102–104]. The expression of many of these factors is also activated during the larval stages and is mediated by a heterogeneous set of specialized immune cells [105, 106]. The adaptive behavior of the powerful sea urchin immune cells and the modifications of their parameters have been recently used as tool for monitoring disease susceptibility and to establish water quality standards and for nano-safety/nano-toxicity analyses [107–109]. In comprehensive review articles about the use of sea urchin embryos for assessing the toxicity of environmental contaminants, the chemical and physical effects of stress during development or in adults are reported in a thematic book focused on the molecular marine biotechnology of Echinoderms [110].

Methods for assessing pharmaceutical and pollutant embryo toxicity and teratogenicity in sea urchin embryos have been developed in sea urchins, indicating that gametes and embryos function as sensitive indicators of environmental toxicity and mutagenicity. Several biological endpoints can be simultaneously evaluated [111, 112], and the long history of sea urchin developmental biology has enormously contributed to these studies.

Teratogenic effects of lithium on sea urchin embryos were first reported more than one century ago [113]. After treatment with LiCl, sea urchin embryos developed as exogastrulae, characterized by an increase in endoderm and mesoderm tissues at the expense of ectoderm. The most visible sign was the alteration in the balance between animal and vegetal regions, with an abnormal and exaggerated presence of vegetally derived cells [114, 115]. It was later shown that embryos cultured in high Li+ concentration delayed the cell cycle, arresting cells at metaphase and at cytokinesis, and that lithium affected the phosphoinositide cycle [116, 117]; this block was, in fact, reversed either after removing lithium or by counteracting its effects by adding myoinositol [118, 119]. Molecular analyses have added information on the properties of lithium, which shift the border between the vegetal and animal embryonic regions. Lithium dramatically increased the expression of vegetal-specific molecular markers [120], even in isolated animal caps [121], and restricted the expression domain of the hatching enzyme toward the animal pole [122, 123]. Lithium, which has been successfully employed in the treatment of bipolar disorders [124], mimicked the activation of Wnt/beta-catenin signaling, producing embryonic phenotypes similar to those elicited by GSK3β inhibition [80]. More recently, it has been reported that sea urchin female gametes had enhanced sensitivity and embryo development was more affected when LiCl was applied before rather than after fertilization [30].

The metallic element nickel perturbs the sea urchin embryo dorsoventral axis, altering the commitment of ectodermal cells. Embryos treated with $NiCl_2$ showed an overrepresentation of oral (ventral) ectoderm cells, causing an increase in the expression of the oral ectoderm-specific markers EctoV and *Orthopedia* [125, 126].

The effects of specific inhibition of the epidermal growth factor receptor (EGFr) are another example of the strength of sea urchin embryonic assays to analyze perturbations at cellular and molecular levels. EGFr inhibition by Tyrphostin AG1478 determined the decrease of mesoderm and endoderm marker expression, the ectopic distribution of the ectoderm-specific hatching enzyme, and the reduced level of β-catenin nuclearization; effects on development were mediated by the MAPK-ERK signaling pathway. The addition of TGF-alpha ligand of EGFr to AG1478-treated embryos completely rescued the embryonic phenotypes either at early or late developmental stages, indicating that the effects of this compound were EGFr-specific (**Figure 2**). Furthermore, AG1478 negatively controlled the expression of the EMT-linked transcription factors Snail and ScratchX/Snai2 and inhibited primary mesenchyme cell specification and EMT [31, 127].

The sea urchin embryo has also been used as model for the assessment of antiproliferative, anti-mitotic, and cytotoxic effects of small molecules. Due to the structural and functional similarities of tubulins among species, including sea urchins [128–130], a number of chemical compounds

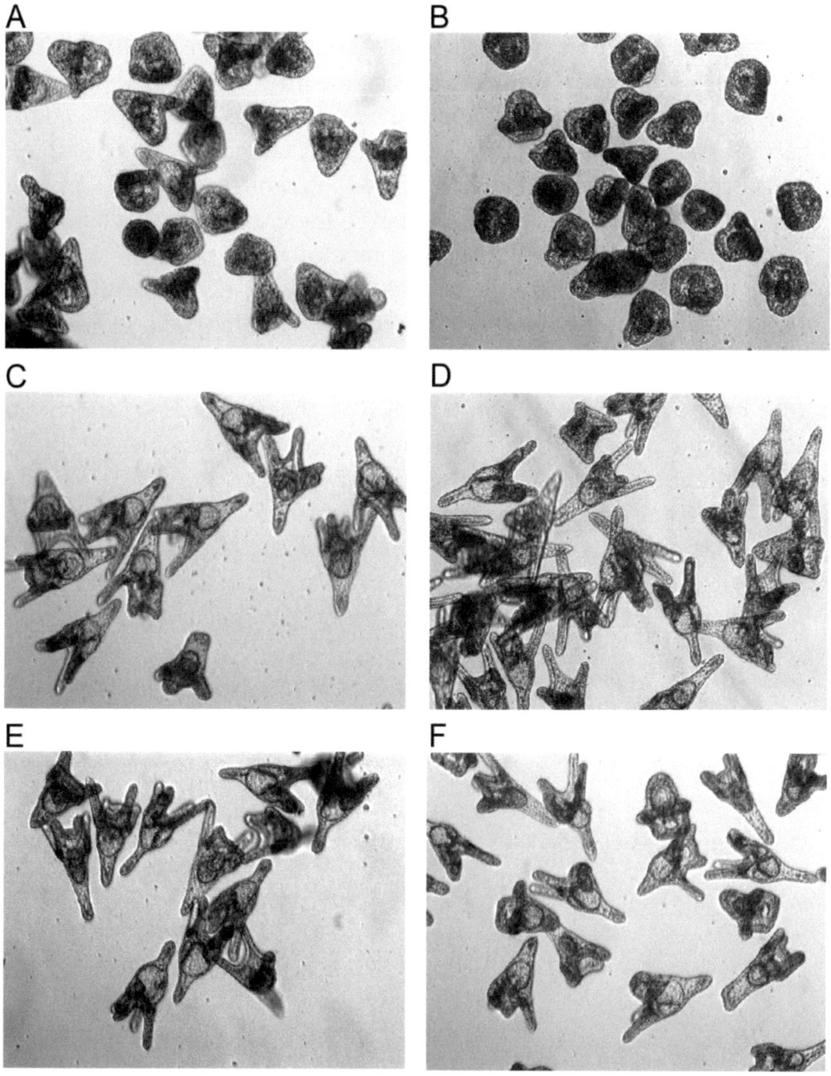

Figure 2. Effects of EGFr inhibition and developmental rescue by TGFα. (A,B) 48 h pluteus stage embryos treated with 10 μg/ml AG1478 from fertilization or from hatching blastula stage, respectively. (C, D) Embryos co-treated with 10 μg/ml AG1478 and 80 ng/ml TGFα. (E) Plutei after treatment with 80 ng/ml TGFα; (F) control plutei.

have been screened for their antimitotic and microtubule-destabilizing activity. The specific effects on embryo development and the swimming behavior of *P. lividus* embryos allowed the identification of chemical agents with specific tubulin destabilizing effects [131–134].

The sea urchin embryo, from fertilization to late embryonic stages, was the subject of several assays aiming to test antitumoral and antiepileptic compounds such as doxorubicin, retinoic acid, phenytoin, valproate, and tamoxifen. Straightforward, reproducible morphological analyses demonstrated that sea urchin embryos function as a simple, specific, and sensitive biological factory for developmental toxicology experiments [135–138]. Reports describing the toxic effects of chemicals, pharmaceuticals, insecticides, pesticides, and other environmental contaminants on sea urchin embryos and adults have been recently published [112, 139–143]. Investigators in marine research have paid particular attention to the fast growth of submicroscopic materials contaminating marine and coastal ecosystems and have analyzed biochemical and histochemical markers of toxicity, taking advantage of the sensitivity of sea urchin gametes, embryos, and adult cells [144]. Recently, researchers have developed new computation methods and a new toxicity index, integrating the frequency of abnormal embryos obtained after treatment with marine waters and sediments with the adverse effects observed during development. This approach has been taking advantage of the high number of morphological parameters that can be measured in sea urchins [145].

Finally, sea urchins play an essential role in ecology, being part of marine benthic communities and functioning as grazers and prey [146]. The physiological performance of marine organisms in challenging the natural environment is endangered by old and emerging hazards coming from anthropic activities and climate-induced changes [147]. Worldwide, living sea urchins, as part of trophic cascades and an important fishery resource, may become threatened in the near future. Assessing the adaptive capacity of these organisms, also in the context of multi-stressor effects, to function as early warning sentinels is thus of increasing interest to researchers and marine organizations, for the management of biological resources and ecosystems [148].

Acknowledgements

Mr. Alessandro Pensato is acknowledged for his skillful support in organizing and assembling images.

Author details

Maria Di Bernardo and Marta Di Carlo*

*Address all correspondence to: marta.dicarlo@ibim.cnr.it

Institute of Biomedicine and Molecular Immunology "A. Monroy" (IBIM), National Research Council, Palermo, Italy

References

[1] Horstadius S. The mechanics of sea urchin development, studied by operative methods. Biological Reviews. 1939;14:132-179

[2] Giudice G. Reconstitution of whole larvae from disaggregated cells of sea urchin embryos. Developmental Biology. 1962;5:402-411

[3] McClay DR, Fink RD. Sea urchin hyalin: Appearance and function in development. Developmental Biology. 1982;92:285-293

[4] Evans T, Rosenthal ET, Youngblom J, Distel, D, Hunt T. Cyclin: A protein specified by maternal mRNA in sea urchin eggs that is destroyed at each cleavage division. Cell. 1983;33:389-396. DOI: 10.1016/0092-8674(83)90420-8

[5] Whitaker M. Calcium at fertilization and in early development. Physiological Reviews. 2006;86:25-88. DOI: http://dx.doi.org/10.1152/physrev.00023.2005

[6] Briggs E, Wessel GM. In the beginning…animal fertilization and sea urchin development. Developmental Biology. 2006;300:15-26. DOI: 10.1016/j.ydbio.2006.07.014

[7] Giudice G. Developmental Biology of the Sea Urchin Embryo. New York: Academic Press; 1973

[8] Chiarelli R, Martino C, Agnello M, Bosco L, Roccheri MC. Autophagy as a defense strategy against stress: Focus on *Paracentrotus lividus* sea urchin embryos exposed to cadmium. Cell Stress & Chaperones. 2016;21:19-27. DOI: 10.1007/s12192-015-0639-3

[9] Di Carlo M, Romancino DP, Montana G, Ghersi G. Spatial distribution of two maternal messengers in *Paracentrotus lividus* during oogenesis and embryogenesis. Proceedings of the National Academy of Sciences of the United States of America. 1994;91:5622-5626

[10] Di Carlo M, Romancino DP, Ortolani G, Montana G, Giudice G. Bep RNAs and proteins are situated in the animal side of sea urchin unfertilized egg, which can be recognized by the female pronuclear location. Biochemical and Biophysical Research Communications. 1996;229:511-517

[11] Romancino DP, Montana G, Di Carlo M. Involvement of the cytoskeleton in localization of *Paracentrotus lividus* maternal BEP mRNAs and proteins. Experimental Cell Research. 1998;238:101-109

[12] Romancino DP, Montana G, Dalmazio S, Di Carlo M. Bep4 protein is involved in patterning along the animal-vegetal axis in the *Paracentrotus lividus* embryos. Developmental Biology. 2001;234:107-119

[13] Giudice G, Roccheri MC, Di Bernardo MG. Synthesis of heat shock proteins in sea urchin embryos. Cell Biology International Reports. 1980;4:69-74

[14] Roccheri MC, Di Bernardo MG, Giudice G. Synthesis of heat shock proteins in developing sea urchins. Developmental Biology. 1981;83:173-177

[15] Roccheri MC, Sconzo G, Di Carlo M, Di Bernardo MG, Pirrone AM, Gambino R, Giudice G. Heat shock proteins in sea urchin embryos. Transcriptional and post-transcriptional regulation. Differentiation. 1982;22:175-178

[16] Agnello M, Roccheri MC. Apoptosis: Focus on sea urchin development. Apoptosis. 2010;15:322-330. DOI: 10.1007/s10495-009-0420-0

[17] Agnello M, Chiarelli R, Martino C, Bosco L, Roccheri MC. Autophagy is required for sea urchin oogenesis and early development. Zygote. 2016;24:918-926. DOI: 10.1017/S0967199416000253

[18] Davidson EH. The sea urchin genome, where would it lead us? Science. 2006;314:939-940. DOI: 10.1126/science.1136252

[19] Sodergren E, Weinstock GM, Davidson EH et al. The genome of the sea urchin *Strongylocentrotus purpuratus*. Science. 2006;314:941-952. DOI: 10.1126/science.1133609

[20] Nakajima Y, Kaneko H, Murray G, Burke RD. Divergent patterns of neural development in larval echinoids and asteroids. Evolution & Development. 2004;6:95-104. DOI: 10.1111/j.1525-142X.2004.04011.x

[21] Hardy J, Selkoe DJ. The amyloid hypothesis of Alzheimer's disease: Progress and problems on the road to therapeutics. Science. 2002;19:353-356. DOI: 10.1126/science.1072994

[22] Pellicanò M, Picone P, Cavalieri V, Carrotta R, Spinelli G, Di Carlo M. The sea urchin embryo: A model to study Alzheimer's beta amyloid induced toxicity. Archives of Biochemistry and Biophysics. 2009;483:120-126. DOI: 10.1016/j.abb.2008.12.006

[23] Di Carlo M. Simple model systems: A challenge for Alzheimer's disease. Immunity and Ageing. 2012;9:3-11. DOI: 10.1186/1742-4933-9-3

[24] Weinert BT, Timiras PS. Theories of aging. Journal of Applied Physiology. 2003;95: 1706-1716

[25] Ebert TA, Southon JR. Red sea urchins (*Strongylocentrotus franciscanus*) can live over 100 years: Confirmation with A-bomb 14carbon. Fishery Bulletin. 2003;101:915-922

[26] Beddingfield SD, McClintock JB. Demographic characteristics of *Lytechinus variegatus* (Echinoidea: Echinodermata) from three habitats in North Florida Bay, Gulf of Mexico. Marine Ecology. 2000;21:17-40

[27] Ebert TA. Demographic patterns of the purple sea urchin *Strongylocentrotus purpuratus* along a latitudinal gradient, 1985-1987. Marine Ecology Progress Series. 2010;406:105-120

[28] Liang J, Aleksanyan H, Metzenberg S, Oppenheimer SB. Involvement of l(-)-rhamnose in sea urchin gastrulation. Part II: α-l-Rhamnosidase. Zygote. 2015;24:371-377. DOI: 10.1017/S0967199415000283

[29] Aleksanyan H, Liang J, Metzenberg S, Oppenheimer SB. Terminal alpha-D-mannosides are critical during sea urchin gastrulation. Zygote. 2016;24:775-782. DOI: 10.1017/S0967199416000113

[30] Ruocco N, Costantini M, Santella L. New insights into negative effects of lithium on sea urchin *Paracentrotus lividus* embryos. Scientific Reports. 2016;6:32157. DOI: 10.1038/srep32157

[31] Romancino D, Anello L, Lavanco A, Buffa V, Di Bernardo M, Bongiovanni A. A sea urchin *in vivo* model to evaluate Epithelial-Mesenchymal transition. Development, Growth & Differentiation. 2017;59:141-151 DOI: 10.1111/dgd.12353

[32] Soto C. Unfolding the role of protein misfolding in neurodegenerative diseases. Nature Reviews Neuroscience. 2003;4:49-60. DOI: 10.1038/nrn1007

[33] Glabe CG, Kayed R. Common structure and toxic function of amyloid oligomers implies a common mechanism of pathogenesis. Neurology. 2006;66(Suppl 1):S74-S78. DOI: 10.1212/01.wnl.0000192103.24796.42

[34] Haass C, Selkoe DJ. Soluble protein oligomers in neurodegeneration: Lessons from the Alzheimer's amyloid beta-peptide. Nature Reviews Molecular Cell Biology. 2007;8:101-112. DOI: 10.1038/nrm2101

[35] Gong Y, Chang L, Viola KL, Lacor PN, Lambert MP, Finch CE, Krafft GA, Klein WL. Alzheimer's disease-affected brain: Presence of oligomeric Abeta ligands (ADDLs) suggests a molecular basis for reversible memory loss. Proceedings of the National Academy of Sciences of the United States of America. 2003;100:10417-10422. DOI: 10.1073/pnas.1834302100

[36] Angelini C, Baccetti B, Piomboni P, Trombino S, Aluigi MG, Stringara S, Gallus L, Falugi C. Acetylcholine synthesis and possible functions during sea urchin development. European Journal of Histochemistry. 2004;48:235-243

[37] Carrotta R, Di Carlo M, Manno M, Montana G, Picone P, Romancino DP, San Biagio PL. Toxicity of recombinant beta-amyloid prefibrillar oligomers on the morphogenesis of the sea urchin *Paracentrotus lividus*. The FASEB Journal. 2006;20:1916-1927. DOI: 10.1096/fj.06-5716fje

[38] Buznikov GA, Nikitina LA, Bezuglov VV, Milosevic I, Lazarevic L, Rogac, S. Ruzdijic L, Slotkin TA, Rakic LM. Sea urchin embryonic development provides a model for evaluating therapies against beta-amyloid toxicity. Brain Research Bulletin. 2008;75:94-100. DOI: 10.1016/j.brainresbull.2007.07.026

[39] Buznikov GA, Nikitina LA, Seidler FJ, Slotkin TA, Bezuglov VV, Milosević I, Lazarević L, Rogac L, Ruzdijić S, Rakić LM. Amyloid precursor protein 96-110 and beta-amyloid 1-42 elicit developmental anomalies in sea urchin embryos and larvae that are alleviated by neurotransmitter analogs for acetylcholine, serotonin and cannabinoids. Neurotoxicology and Teratology. 2008;30:503-509. DOI: 10.1016/j.ntt.2008.05.003

[40] Small DH, Nurcombe V, Reed G, Clarris H, Moir R, Beyreuther K, Masters CL. A heparin-binding domain in the amyloid protein precursor of Alzheimer's disease is involved in the regulation of neurite outgrowth. The Journal of Neuroscience. 1994;14:2117-2127

[41] Rosen DR, Martin-Morris L, Luo LQ, White K. A Drosophila gene encoding a protein resembling the human beta-amyloid protein precursor. Proceedings of the National Academy of Sciences of the United States of America. 1989;86:2478-2482

[42] Daigle I, Li C. Apl-1, a *Caenorhabditis elegans* gene encoding a protein related to the human beta-amyloid protein precursor. Proceedings of the National Academy of Sciences of the United States of America. 1993;90:12045-12049

[43] Coulson EJ, Paliga K, Beyreuther K, Masters CL. What the evolution of the amyloid protein precursor supergene family tells us about its function. Neurochemistry International. 2000;36:175-184. DOI: 10.1016/S0197-0186(99)00125-4

[44] Dickson DW. Apoptotic mechanisms in Alzheimer neurofibrillary degeneration: Cause or effect? The Journal of Clinical Investigation. 2004;114:121-130. DOI: 10.1172/JCI200420640

[45] Wang KK. Calpain and caspase: Can you tell the difference? Trends in Neurosciences. 2000;23:20-26. DOI: 10.1016/S0166-2236(99)01536-2

[46] Vega Thurber R, Epel D. Apoptosis in early development of the sea urchin, *Strongylocentrotus purpuratus*. Developmental Biology. 2007;303:336-346. DOI: 10.1016/j.ydbio.2006.11.018

[47] Crescenzi O, Tomaselli S, Guerrini R, Salvatori S, D'Ursi AM, Temussi PA, Picone D. Solution structure of the Alzheimer amyloid beta-peptide (1-42) in an apolar microenvironment. Similarity with a virus fusion domain. European Journal of Biochemistry. 2002;269:5642-5648. DOI: 10.1021/bi060998w

[48] Govindarajua D, Atzmonb G, Barzilai N. Genetics lifestyle and longevity: Lessons from centenarians. Applied & Translational Genomics. 2015;4:23-32. DOI: 10.1016/j.atg.2015.01.001

[49] Vasto S, Buscemi S, Barera A, Di Carlo M, Accardi G, Caruso C. Mediterranean diet and healthy ageing: A sicilian perspective. Gerontology. 2014;60:508-518. DOI: 10.1159/000363060

[50] Vasto S, Barera A, Rizzo C, Di Carlo M, Caruso C, Panotopoulos G. Mediterranean diet and longevity: An example of nutraceuticals? Current Vascular Pharmacology. 2014;12:735-738

[51] Finch CE, Austad SN. History and prospects: Symposium on organisms with slow aging. Experimental Gerontology. 2001;36:593-597. DOI: 10.1016/S0531-5565(00)00228-X

[52] Bodnar A. Cellular and molecular mechanisms of negligible senescence: Insight from the sea urchin. Invertebrate Reproduction and Development. 2015;59:23-27. DOI: 10.1080/07924259.2014.938195

[53] Ebert TA Longevity and lack of senescence in the red sea urchin *Strongylocentrotus* franciscanus. Experimental Gerontology. 2008;43:734-738. DOI: 10.1016/j.exger.2008.04.015

[54] Kobayashi S, Taki J. Calcification in sea urchins. A tetracycline investigation of growth of the mature test in *Strongylocentrotus* intermedius. Calcified Tissue Research. 1969;4: 210-223

[55] Kirkwood TB. Evolution of ageing. Mechanisms of Ageing and Development. 2002;123:737-745. DOI: 10.1016/S0047-6374(01)00419-5

[56] Bodnar A. Proteomic profiles reveal age-related changes in coelomic fluid of sea urchin species with different life spans. Experimental Gerontology. 2013;48:525-530. DOI: 10.1016/j.exger.2013.01.014

[57] Du C, Anderson A, Lortie M, Parson R, Bodnar A. Oxidative damage and cellular defense mechanisms in sea urchin models of aging. Free Radical Biology and Medicine. 2013;63:254-263. DOI: 10.1016/j.freeradbiomed.2013.05.023

[58] Graf E. Antioxidant potential of ferulic acid. Free Radical Biology and Medicine. 1992;13:435-448. DOI: 10.1016/0891-5849(92)90184-I

[59] Ozaki Y. Antiinflammatory effect of tetramethylpyrazine and ferulic acid. Chemical & Pharmaceutical Bulletin. 1992;40:954-956

[60] Kanski J, Aksenova M, Stoyanova A, Butterfield DA. Ferulic acid antioxidant protection against hydroxyl and peroxyl radical oxidation in synaptosomal and neuronal cell culture systems *in vitro*: Structure-activity studies. The Journal of Nutritional Biochemistry. 2002;13:273-281. DOI: 10.1016/S0955-2863(01)00215-7

[61] Picone P, Bondi ML, Montana G, Bruno A, Pitarresi G, Giammona G, Di Carlo M. Ferulic acid inhibits oxidative stress and cell death induced by Ab oligomers: Improved delivery by solid lipid nanoparticles. Free Radical Research. 2009;43:1133-1145. DOI: 10.1080/10715760903214454

[62] Sgarbossa A, Giacomazza D, Di Carlo M. Ferulic acid: A hope for Alzheimer's disease therapy for plants. Nutrients. 2015;7:5764-5782. DOI: 10.3390/nu7075246

[63] Picone P, Nuzzo D, Di Carlo M. Ferulic acid: A natural antioxidant against oxidative stress induced by oligomeric Abeta on sea urchin embryo. The Biological Bulletin. 2013;224:18-28. DOI: 10.1086/BBLv224n1p18

[64] Sea Urchin Genome Sequencing Consortium. The genome of the sea urchin *Strongylocentrotus* purpuratus. Science. 2006;314:941-952

[65] Cameron RA, Davidson E. A basal deuterostome genome viewed as a natural experiment. Gene. 2007;406:1-7. DOI: 10.1016/j.gene.2007.04.031

[66] Cameron RA, Kudtarkar P, Gordona SM, Worley KC, Gibbs RA. Do echinoderm genomes measure up? Marine Genomics. 2015;22:1-9. DOI: 10.1016/j.margen.2015.02.004

[67] Davidson EH, Rast JP, Oliveri P, Ransick A, Calestani C, Yuh CH, Minokawa T, Amore G, Hinman V, Arenas-Mena C, et al. A genomic regulatory network for development. Science. 2002;295:1669-1678. DOI: 10.1126/science.1069883

[68] Davidson EH, Erwin DH. Gene regulatory networks and the evolution of animal body plans. Science. 2006;311:796-800. DOI: 10.1126/science.1113832

[69] Arnone MI, Andrikou C, Annunziata R. Echinoderm systems for gene regulatory studies in evolution and development. Current Opinion in Genetics and Development. 2016;39:129-137. DOI: 10.1016/j.gde.2016.05.027

[70] Cary GA, Hinman VF. Echinoderm development and evolution in the post-genomic era. Developmental Biology. 2017;427:203-211. DOI: 10.1016/j.ydbio.2017.02.003

[71] Peter IS, Davidson EH. Evolution of gene regulatory networks controlling body plan development. Cell. 2011;44:970-985. DOI: 10.1016/j.cell.2011.02.017

[72] Wu S-Y, Ferkowicz M, McClay DR. Ingression of primary mesenchyme cells of the sea urchin embryo: A precisely timed epithelial mesenchymal transition. Birth Defects Research (Part C). 2007;81:241-252. DOI: 10.1002/bdrc.20113

[73] Katow H. Mechanisms of the epithelial-to-mesenchymal transition in sea urchin embryos. Tissue Barriers. 2015;3(4):e1059004. DOI: 10.1080/21688370.2015.1059004

[74] Lim J, Thiery JP. Epithelial-mesenchymal transitions: Insights from development. Development. 2012;139:3471-3486. DOI: 10.1242/dev.071209

[75] Nieto MA, Huang RY-J, Jackson RA, Thiery JP. EMT. Cell. 2016;166:21-45. DOI: 10.1016/j.cell.2016.06.028

[76] Wu CY, Tsai YP, Wu MZ, Teng SC, Wu KJ. Epigenetic reprogramming and post-transcriptional regulation during the epithelial-mesenchymal transition. Trends in Genetics. 2012;28:454-463. DOI: 10.1016/j.tig.2012.05.005

[77] Saunders LR, McClay DR. Sub-circuits of a gene regulatory network control a developmental epithelial-mesenchymal transition. Development. 2014;141:1503-1513. DOI: 10.1242/dev.101436

[78] Cavalieri V, Spinelli G. Early asymmetric cues triggering the dorsal/ventral gene regulatory network of the sea urchin embryo. eLife. 2014;3:e04664. DOI: 10.7554/eLife.04664

[79] Croce JC, McClay DR. The canonical Wnt pathway in embryonic axis polarity. Seminars in Cell & Developmental Biology. 2006;17:168-174. DOI: 10.1016/j.semcdb.2006.04.004

[80] Emily-Fenouil F, Ghiglione C, Lhomond G, Lepage T, Gache C. GSK3b/shaggy mediates patterning along the animal-vegetal axis of the sea urchin embryo. Development. 1998;125:2489-2498

[81] Fernandez-Serra M, Consales C, Livigni A, Arnone MI. Role of the ERK-mediated signaling pathway in mesenchyme formation and differentiation in the sea urchin embryo. Developmental Biology. 2004;268:384-402. DOI: 10.1016/j.ydbio.2003.12.029

[82] Logan CY, Miller JR, Ferkowicz MJ, McClay DR. Nuclear β-catenin is required to specify vegetal cell fates in the sea urchin embryo. Development. 1999;126:345-357

[83] Molina MD, de Crozé N, Haillot E, Lepage T. Nodal: Master and commander of the dorsal-ventral and left-right axes in the sea urchin embryo. Current Opinion in Genetics & Development. 2013;23:445-453. DOI: 10.1016/j.gde.2013.04.010

[84] Röttinger E, Besnardeau L, Lepage T. A Raf/MEK/ERK signaling pathway is required for development of the sea urchin embryo micromere lineage through phosphorylation of the transcription factor Ets. Development. 2004;131:2233. DOI: 10.1242/dev.01000

[85] Weitzel HE, Illies MR, Byrum CA, Xu R, Wikramanayake AH, Ettensohn CA. Differential stability of beta-catenin along the animal-vegetal axis of the sea urchin embryo mediated by dishevelled. Development. 2004;131:2947-2956. DOI: 10.1242/dev.01152

[86] Wu SY, McClay DR. The Snail repressor is required for PMC ingression in the sea urchin embryo. Development. 2007;134:1061-1070. DOI: 10.1242/dev.02805

[87] Ettensohn CA, Wessel GM, Wray GA, editors. Development of Sea Urchins, Ascidians, and Other Invertebrate Deuterostomes: Experimental Approaches. Meth Cell Biol. 74. Elsevier Academic press, Amsterdam;2004. p. 875. ISBN 0-12-480278-8

[88] Goldstone JV, Hamdoun A, Cole BJ. Howard-Ashby M, Nebert DW, Scally M, Dean M, Epel D, Hahn ME, Stegeman JJ. The chemical defensome: Environmental sensing and response genes in the *Strongylocentrotus* purpuratus genome Developmental Biology. 2006;300:366-384. DOI: 10.1016/j.ydbio.2006.08.066

[89] Rast JP, Smith LC, Loza-Coll M, Hibino T, Litman GW. Genomic insights into the immune system of the sea urchin. Science. 2006;314:952-956. DOI: 10.1126/science.1134301

[90] Dean M, Hamon Y, Chimini G. The human ATP-binding cassette (ABC) transporter superfamily. Journal of Lipid Research. 2001;42:1007-1017

[91] Hamdoun AM, Cherr GN, Roepke TA, Epel D. Activation of multidrug efflux transporter activity at fertilization in sea urchin embryos (*Strongylocentrotus purpuratus*). Developmental Biology. 2004;276:452-462. DOI: 10.1016/j.ydbio.2004.09.013

[92] Epel D, Cole B, Hamdoun A, Vega Thurber R. The sea urchin embryo as a model for studying efflux transporters: Roles and energy cost. Marine Environmental Research. 2006;62:S1-S4. DOI: 10.1016/j.marenvres.2006.04.062

[93] Cole BJ, Hamdoun A, Epel D. Cost, effectiveness and environmental relevance of multidrug transporters in sea urchin embryos. Journal of Experimental Biology. 2013;216:3896-3905. DOI: 10.1242/jeb.090522

[94] Bonacci S, Iacocca A, Fossi S, Lancini L, Caruso T, Corsi I, Focardi S. Biomonitoring aquatic environmental quality in a marine protected area: A biomarker approach. Ambio. 2007;36:308-315

[95] Geraci F, Pinsino A, Turturici G, Savona R, Giudice G, Sconzo G. Nickel, lead, and cadmium induce differential cellular responses in sea urchin embryos by activating the synthesis of different HSP70s. Biochemical and Biophysical Research Communications. 2004;322:873-877. DOI: 10.1016/j.bbrc.2004.08.005

[96] Bonaventura R, Poma V, Costa C, Matranga V. UVB radiation prevents skeleton growth and stimulates the expression of stress markers in sea urchin embryos. Biochemical and Biophysical Research Communications. 2005;328:150-157. DOI: 10.1016/j.bbrc.2004.12.161

[97] Pinsino A, Matranga V. Sea urchin immune cells as sentinels of environmental stress. Developmental and Comparative Immunology. 2005;49:198-205

[98] Muralidharan S, Mandrekar P. Cellular stress response and innate immune signaling: Integrating pathways in host defense and inflammation. Journal of Leukocyte Biology. 2013;94:1167-1184. DOI: 10.1189/jlb.0313153

[99] Smith LC. Diversification of innate immune genes: Lessons from the purple sea urchin. Disease Models and Mechanisms. 2010;3:274-279. DOI: 10.1242/dmm.004697

[100] Buckley KM, Rast JP. Diversity of animal immune receptors and the origins of recognition complexity in the deuterostomes. Developmental and Comparative Immunology. 2015;49:179-189. DOI: 10.1016/j.dci.2014.10.013

[101] Motta V, Soares F, Sun T, Philpott DJ. NOD-like receptors: Versatile cytosolic sentinels. Physiological Reviews. 2015;95:149-178. DOI: 10.1152/physrev.00009.2014

[102] Hibino T, Loza-Coll M, Messier C, Majeske AJ, Cohen AH, Terwilliger DP, Buckley KM, Brockton V, Nair SV, Berney K, Fugmann SD, Anderson MK, Pancer Z, Cameron RA, Smith LC, Rast JP. The immune gene repertoire encoded in the purple sea urchin genome. Developmental Biology. 2006;300:349-365. DOI: 10.1016/j.ydbio.2006.08.065

[103] Ghosh J, Buckley KM, Nair SV, Raftos DA, Miller C, Majeske AJ, Hibino T, Rast JP, Roth M, Smith LC. Sp185/333: A novel family of genes and proteins involved in the purple sea urchin immune response. Developmental and Comparative Immunology. 2010;34:235-245. DOI: 10.1016/j.dci.2009.10.008

[104] Sherman LS, Schrankel CS, Brown KJ, Smith LC. Extraordinary diversity of immune response proteins among sea urchins: Nickel-isolated Sp185/333 proteins show broad variations in size and charge. PloS One. 2015;10:e0138892. DOI: 10.1371/journal.pone.0138892

[105] Hirano M. Dynamic immune response in echinoderm larvae Echinoderm immunity: Is the larval immune system immature? Immunology and Cell Biology. 2016;94:809-811. DOI: 10.1038/icb.2016.67

[106] Ho ECH, Buckley KM, Schrankel CS, Schuh NW, Hibino T Solek CM, Bae K, Wang G, Rast JP. Perturbation of gut bacteria induces a coordinated cellular immune response in the purple sea urchin larva. Immunology and Cell Biology. 2016;94:861-874. DOI: 10.1038/icb.2016.51

[107] Manzo S, Schiavo S, Oliviero M, Toscano A, Ciaravolo M, Cirino P. Immune and reproductive system impairment in adult sea urchin exposed to nanosized ZnO via food. Science of the Total Environment. 2017;599-600:9-13. DOI: 10.1016/j.scitotenv.2017.04.173

[108] Pinsino A, Russo R, Bonaventura R, Brunelli A, Marcomini A, Matranga V. Titanium dioxide nanoparticles stimulate sea urchin immune cell phagocytic activity involving

TLR/p38 MAPK-mediated signalling pathway. Scientific Reports. 2015;5:14492. DOI: 10.1038/srep14492

[109] Stabili S, Pagliara P. The sea urchin *Paracentrotus lividus* immunological response to chemical pollution exposure: The case of lindane. Chemosphere. 2015;134:60-66. DOI: 10.1016/j.chemosphere.2015.04.006

[110] Matranga, editor. Echinodermata series: Progress in molecular and subcellular biology. In: Werner E, Müller G, editor. Subseries: Marine Molecular Biotechnology. Vol. 39. Springer-Verlag Berlin 2005. ISSN 1611-6119

[111] Hose JE. Potential uses of sea urchin embryos for identifying toxic chemicals: Description of a bioassay incorporating cytologic, cytogenetic and embryologic endpoints. Journal of Applied Toxicology. 1985;5:245-254

[112] Ribeiro S, Torres T, Martins R, Santos MM. Toxicity screening of diclofenac, propranolol, sertraline and simvastatin using *Danio rerio* and *Paracentrotus lividus* embryo bioassays. Ecotoxicology and Environmental Safety. 2015;114:67-74. DOI: 10.1016/j.ecoenv.2015.01.008

[113] Herbst C. Experimentelle untersuchungen iiber den einfluss der veranderten chemischen zusammensetzung des umgebenden mediums auf die entwicklung der tiere. I teil Versuche an Seeigeleiern.. Zeitschrift für wissenschaftliche Zoologie. 1982;55:446-518

[114] Lallier R. Biochemical aspects of animalization and vegetalization in the sea urchin embryo. Advances in Morphogenesis. 1964;3:147-196

[115] Horstadius S. Experimental Embryology of Echinoderms. Oxford: Clarendon Press; 1973

[116] Berridge MJ, Downes CP, Hanley MR. Neural and developmental actions of lithium: A unifying hypothesis. Cell. 1989;59:411-419. DOI: 10.1016/0092-8674(89)90026-3

[117] Ciapa B, Maggio K. Effect of lithium on ionic balance and polyphosphoinositide metabolism during larval vegetalization of the sea urchin *Paracentrotus lividus*. Developmental Biology. 1993;159:114-121. DOI: 10.1006/dbio.1993.1225

[118] Becchetti A, Whitaker M. Lithium blocks cell cycle transitions in the first cell cycles of sea urchin embryos, an effect rescued by myo-inositol. Development. 1997;124:1099-1107

[119] Sconzo G, Cascino D, Amore G, Geraci, F, Giudice G. Effect of the IMPase inhibitor L690,330 on sea urchin development. Cell Biology International. 1998;22:91-94

[120] Nocente-McGrath C, McIsaac R, Ernst SG. Altered cell fate in LiCl-treated sea urchin embryos. Developmental Biology. 1991;147:445-450. DOI: 10.1016/0012-1606(91)90302-J

[121] Livingston BT, Wilt FH. Lithium evokes expression of vegetal-specific molecules in the animal blastomeres of sea urchin embryos. Proceedings of the National Academy of Sciences of the United States of America. 1989;86:3669-3673

[122] Ghiglione C, Lhomond G, Lepage T, Gache C. Cell-autonomous expression and position-dependent repression by Li+ of two zygotic genes during sea urchin early development. The EMBO Journal. 1993;12:87-96

[123] Ghiglione C, Emily-Fenouil F, Chang P, Gache C. Early gene expression along the animal-vegetal axis in sea urchin embryoids and grafted embryos. Development. 1996; 122:3067-3074

[124] Malhi GS, Tanious M, Das P, Coulston CM, Berk M. Potential mechanisms of action of lithium in bipolar disorder. Current understanding. CNS Drugs. 2013;27:135-153. DOI: 10.1007/s40263-013-0039-0

[125] Di Bernardo M, Castagnetti S, Bellomonte D, Oliveri P, Melfi R, Palla F, Spinelli G. Spatially restricted expression of PlOtp, a Paracentrotus lividus Orthopedia related homeobox gene, is correlated with oral ectodermal patterning and skeletal morphogenesis in late-cleavage sea urchin embryos. Development. 1999;126:2171-2179

[126] Hardin J, Coffman JA, Black SD, McClay DR. Commitment along the dorsoventral axis of the sea urchin embryo is altered in response to $NiCl_2$. Development. 1992;116:671-685

[127] Romancino DP, Montana G, Cavalieri V, Spinelli G, Di Carlo M. EGFR signalling is required for *Paracentrotus lividus* endomesoderm specification. Archives of Biochemistry and Biophysics. 2008;474:167-174. DOI: 10.1016/j.abb.2008.03.022

[128] Cleveland DW. The multitubulin hypothesis revisited: What have we learned? The Journal of Cell Biology. 1987;104:381-383. DOI: 0021-9525/87/03/381/3

[129] Gianguzza F, Di Bernardo MG, Sollazzo M, Palla F, Ciaccio M, Carra E, Spinelli G. DNA sequence and pattern of expression of the sea urchin (*Paracentrotus lividus*) α-tubulin genes. Molecular Reproduction and Development. 1989;1:170-181. DOI: 10.1002/mrd.1080010305

[130] McKean PG, Vaughan S, Gull K. The extended tubulin superfamily Journal of Cell Science. 2011;114:2723-2733

[131] Eurtivong C, Semenov V, Semenova M, Konyushkin L, Atamanenko O, Reynisson J, Kiselyov A. 3-Amino-thieno[2,3-b]pyridines as microtubule-destabilising agents: Molecular modelling and biological evaluation in the sea urchin embryo and human cancer cells. Bioorganic and Medicinal Chemistry. 2017;25:658-664. DOI: 10.1016/j.bmc.2016.11.041

[132] Kiselyov AS, Semenova MN, Chernyshova NB, Leitao A, Samet AV, Kislyi KA, Raihstat MM, Oprea T, Lemcke H, Lantow M, Weiss DG, Ikizalp NN, Kuznetsov SA, Semenov VV. Novel derivatives of 1,3,4-oxadiazoles are potent mitostatic agents featuring strong microtubule depolymerizing activity in the sea urchin embryo and cell culture assays. European Journal of Medicinal Chemistry. 2010;45:1683-1697. DOI: 10.1016/j.ejmech.2009.12.072.

[133] Semenov VV, Lichitsky BV, Komogortsev AN, Dudinov AA, Krayushkin MM, Konyushkin LD, Atamanenko OP, Karmanova IB, Strelenko YA, Shor B, Semenova MN, Kiselyov AS. Synthesis and anti-mitotic activity of 6,7-dihydro-4H-isothiazolo[4,5-b]pyridin-5-ones: *In vivo* and cell-based studies. European Journal of Medicinal Chemistry. 2017; 125:573-585. DOI: 10.1016/j.ejmech.2016.09.075

[134] Semenova MN, Kiselyov A, Semenov VV. Sea urchin embryo as a model organism for the rapid functional screening of tubulin modulators BioTechniques. 2006;40:765-774. DOI: 10.2144/000112193

[135] Sconzo G, Romancino D, Fasulo G, Cascino D, Giudice G. Effect of doxorubicin and phenytoin on sea urchin development. Pharmazie. 1995;50:616-619

[136] Sconzo G, Fasulo G, Romancino D, Cascino D, Giudice G. Effect of retinoic acid and valproate on sea urchin development. Pharmazie. 1996;51:175-180

[137] Macedo S, Torres T, Santos MM. Methyl-triclosan and triclosan impact embryonic development of *Danio rerio* and *Paracentrotus lividus*. Ecotoxicology. 2017;26:482-489. DOI: 10.1007/s10646-017-1778-3

[138] Pagano G, de Biase A, Deeva IB, Degan P, Doronin YK, Iaccarino M, Orale R, Trieff NM, Warnaug M, Korkina LG. The role of oxidative stress in developmental and reproductive toxicity of tamoxifen. Life Sciences. 2001;68:1735-1749. DOI: 10.1016/S0024-3205(01)00969-9

[139] Buttino I, Hwang JS, Romano G, Sun CK, Liu TM, Pellegrini D, Gaion A, Sartori D. Detection of malformations in sea urchin plutei exposed to mercuric chloride using different fluorescent techniques. Ecotoxicology and Environmental Safety. 2016;123:72-80. DOI: 10.1016/j.ecoenv.2015.07.027

[140] Cunha I, Torres T, Oliveira H, Martins R, McGowan T, Sheahan D, Machado Santos M. Using early life stages of marine animals to screen the toxicity of priority hazardous and noxious substances. Environmental Science and Pollution Research. 2017;24:10510-10518. DOI: 10.1007/s11356-017-8663-8

[141] Kanold JM, Wang J, Brümmer F, Šiller L. Metallic nickel nanoparticles and their effect on the embryonic development of the sea urchin *Paracentrotus lividus*. Environmental Pollution. 2016;212:224-229. DOI: 10.1016/j.envpol.2016.01.050

[142] Martino C, Bonaventura R, Byrne M, Roccheri M, Matranga V. Effects of exposure to gadolinium on the development of geographically and phylogenetically distant sea urchins species. Marine Environmental Research. 2016;S0141-1136:30098-30108. DOI: 10.1016/j.marenvres.2016.06.001

[143] Motta CM, Cerciello R, De Bonis S, Mazzella V, Cirino P, Panzuto R, Ciaravolo M, Simoniello P, Toscanesi M, Trifuoggi M, Avallone B. Potential toxicity of improperly discarded exhausted photovoltaic cells. Environmental Pollution. 2016;216:786-792. DOI: 10.1016/j.envpol.2016.06.048

[144] Gambardella C, Ferrando S, Gatti AM, Cataldi E, Ramoino P, Aluigi MG, Faimali M, Diaspro A, Falugi C. Morphofunctional and biochemical markers of stress in sea urchin life stages exposed to engineered nanoparticles. Environmental Toxicology. 2015;31:1552-1562. DOI: 10.1002/tox.22159

[145] Morroni L, Pinsino A, Pellegrini D, Regoli F, Matranga V. Development of a new integrative toxicity index based on an improvement of the sea urchin embryo toxicity

test. Ecotoxicology and Environmental Safety. 2016;123:2-7.DOI: 10.1016/j.ecoenv.2015.09.026

[146] Pearse JS. Ecological role of purple sea urchins. Science. 2006;314:940-941. DOI: 10.1126/science.1131888

[147] Byrne M. Global change ecotoxicology: Identification of early life history bottlenecks in marine invertebrates, variable species responses and variable experimental approaches. Marine Environmental Research. 2012;76:3-15. DOI: 10.1016/j.marenvres.2011.10.004

[148] Chan F, Boehm AB, Barth JA, Chornesky EA, Dickson AG, Feely RA, Hales B, Hill TM, Hofmann G, Ianson D, Klinger T, Largier J, Newton J, Pedersen TF, Somero GN, Sutula M, Wakefield WW, Waldbusser GG, Weisberg SB, Whiteman EA. The West Coast Ocean Acidification and Hypoxia Science Panel: Major Findings, Recommendations, and Actions. Oakland, California, USA: California Ocean Science Trust; April 2016